North American Ducks, Geese & Swans Identification Guide

© 2018 Frank S. Todd

All rights reserved. No part of this book may be reproduced in any form whatever without the express consent of Frank S. Todd, except for brief extracts used in reviews or scholarly works.

ISBN Number: ISBN-13: 978-0-692-07902-7

Design and layout by John and Lisa Bass

Printed in Korea

NORTH AMERICAN DUCKS, GEESE & SWANS

IDENTIFICATION GUIDE

Frank S. Todd

TABLE OF CONTENTS

Dedication, Preface and Acknowledgements 1
Whistling-Ducks Introduction 7
  Black-bellied Whistling-Duck 8
  Fulvous Whistling-Duck 10
  West Indian Whistling-Duck 11
Geese Introduction 12
  Greater White-Fronted Goose 13
  Pacific White-Fronted Goose 13
  Tule Goose 15
  Gambel's White-Fronted Goose 16
  Greenland White-Fronted Goose 16
  Lesser White-Fronted Goose 17
  Bean Goose 19
  Pink-Footed Goose 20
  Greylag Goose 21
  Emperor Goose 22
  Snow Goose 24
  Lesser Snow Goose 24
  Blue Goose 25
  Greater Snow Goose 26
  Ross's Goose 27
  Blue Morph Ross's Goose 28
  Canada Goose 29
  Western Canada Goose 29
  Giant Canada Goose 30
  Atlantic Canada Goose 30
  Dusky Canada Goose 31
  Vancouver Canada Goose 31
  Lesser Canada Goose 32
  Interior (Hudson Bay) Canada Goose 32
  Cackling Goose 33
  Richardson's Cackling Goose 33
  Taverner's Cackling Goose 33
  Aleutian Cackling Goose 34
  Hawaiian Goose (Nene) 35
  Barnacle Goose 37
  Brent Goose 38
  Black Brant (Pacific Brent Goose) 38
  Atlantic (Pale-Bellied) Brent 39
Swans Introduction 41
  Trumpeter Swan 42
  Whooper Swan 44
  Tundra Swan 46
  Tundra (Whistling) Swan 46
  Tundra (Bewick's) Swan 48
  Mute Swan 50
Perching-Ducks Introduction 52
  Wood Duck 53
  Mandarin Duck 56
  Muscovy Duck 58
Dabbling-Ducks Introduction 60
  Gadwall 61
  American Wigeon 63
  Eurasian Wigeon 65
  Falcated Duck 67
  Mallard 69
  Greenland Mallard 71
  Mexican Duck 72
  Mariana Mallard 72
  American Black Duck 73
  Mottled Duck 75
  Gulf Mottled Duck 75

Florida Mottled Duck 76
Laysan Duck 77
Hawaiian Duck (Koloa Maoli) 79
Eastern Spot-Billed Duck 81
Blue-Winged Teal 82
Cinnamon Teal 84
Northern Shoveler 86
Northern Pintail 89
White-Cheeked Pintail 91
Garganey 93
Baikal Teal 94
Green-Winged Teal 96
Eurasian Green-Winged Teal 98
Pochards Introduction 99
Canvasback 100
Redhead 103
Common (Eurasian) Pochard 105
Baer's Pochard 106
Ring-Necked Duck 108
Tufted Duck 111
Greater Scaup 113
Lesser Scaup 115
Eiders Introduction 117
Common Eider 118
American Eider 118
Northern Eider 120
Pacific Eider 121
Hudson Bay Eider 122
Spectacled Eider 123
King Eider 126
Steller's Eider 129
Sea-Ducks Introduction 132
Harlequin Duck 133
Labrador Duck 137
Black Scoter 138
Common Scoter 141
Surf Scoter 143
White-Winged Scoter 147
Velvet Scoter 150
Siberian Scoter 150
Long-Tailed Duck (Oldsquaw) 151
Bufflehead 154
Common Goldeneye 157
Barrow's Goldeneye 160
Hooded Merganser 162
Smew 165
Common Merganser 167
Red-Breasted Merganser 171
Stiff-Tailed Ducks Introduction 174
Ruddy Duck 175
Masked Duck 178
Urban Waterfowl 182
Black Swan 183
Swan Goose 184
Bar-Headed Goose 185
Red-Breasted Goose 186
Egyptian Goose 188
Common Shelduck 189
Ruddy Shelduck 190
Ferruginous Duck 191
Red-Crested Pochard 192
Appendix 193
Index 199

Dedication

**To Lissie, a shimmering light
who illuminated the road briefly for those to follow in her footsteps.**

Foreward

In April 2015, our very good friend Frank S. Todd, and a friend to thousands of people around the planet, was diagnosed with skin cancer. He began various treatments for this insidious disease soon after that and by late 2015 it appeared to be under control. But in November 2016 the cancer spread and began working against him, and us all, and on 8 December 2016 Frank moved on.

Throughout those last two years Frank, nonetheless, continued tirelessly to promote conservation and preservation of wildlife (particularly waterfowl and penguins), and their habitats everywhere on earth and to collect materials needed to finish this wonderful *Identification Guide to North American Ducks, Geese & Swans*. Indeed, as always, Frank raged against the dying of the light. He surely did not go gently into the good night and we suspect that his spirit continues to rage for all the things that he was committed to during his physical tenure on Earth. We, at least, can still hear him now and likely always will.

In early 2016 Frank asked us to finish the final stages of his Identification Guide and get it printed in the event that he was not able to. Typical of Frank's serious joking nature, he said "If I die before this book is finished, John is going to kill me". We didn't take this as a joke, but we instead unconditionally agreed to work to get his book finished and printed.

Frank's *Identification Guide to North American Ducks, Geese & Swans* is, we think, another masterpiece in his productions. Long may it run and long may it carry Frank's spirit along with it!

Brent S. Stewart, John Bass & Pamela K. Yochem
South Georgia Island, San Nicolas Island, San Diego (California), Santa Fe (New Mexico)

Preface

Of the more than 10,500 species of birds on Earth, ducks, geese and swans traditionally rank in the top five of the most admired birds, with the "love affair" between humans and them having endured for thousands of years. For purposes of this volume, North America encompasses Mexico, the Caribbean and Greenland, which, at its closest point in the northwest is merely 14 miles from Ellesmere Island, Canada. As the Hawaiian Islands are part of the U.S., they are obviously included, as are U.S. territories in the Pacific (the Marianas Islands including Guam, Saipan, Tinian, Rota, and Aguigan Islands). In dimorphic ducks males and females in breeding plumage are depicted, as well as drake eclipse plumage, transitional plumages, juvenile male and sub-adult plumages, downy young, upper wing pattern, and flight images. Nuptial-plumaged drakes, especially those with iridescent metallic head reflections, can appear as entirely different birds under varying light conditions. Nearly all of the dimorphic northern duck drakes assume an obscure eclipse plumage in summer that recalls the female, during which time all flight feathers are molted, rendering the ducks flightless. The drab female-type plumage evidently affords the flightless males some degree of camouflage that may conceal them from predators. Several distinctive behavioral postures are also included to illustrate the beauty of the various species. Brief bullet points highlight some of the important identification features, but this is primarily a visual book with little text because there are countless books with reams of written descriptions if more detail is sought. Range maps for all species are provided - blue for nesting range, green for wintering ranges or areas of mostly regular dispersal, and red to indicate ranges where permanent and wintering ranges overlap. Forty-four species of waterfowl regularly breed in the mainland United States and Canada. But once the Mexican and Caribbean species, vagrants, and accidentals are counted, there are at least 73 species (125 forms including subspecies) that occur there. The appendix summarizes body masses of both sexes, clutch sizes, incubation times, and fledging periods. Average body mass is given, but many waterfowl fluctuate in mass by as much as 50 percent, especially for highly migratory species. The most recent population status of the majority of each species that breed exclusively in the coverage area are noted and estimates of global population sizes are provided for those species that also breed elsewhere.

Acknowledgments

It would be impossible to even consider a comprehensive volume like this one without considerable assistance. At least 15 years were devoted to this project, with ornithologists and enthusiastic and skilled aviculturists everywhere providing considerable valuable support and direct help. In the very early days, I was also most fortunate to have benefitted greatly from input from and discussions with some of the true ornithological giants of aviculture, including Dr. Jean Delacour, Dr. Paul Johnsgard, Sir Peter Scott and Dr. S. Dillon Ripley, all who greatly encouraged me in countless ways.

For providing grant and sponsorship funds to support my travel to collect photographs and materials to prepare the book, I thank Naidine Adams (*Harry G. and Pauline M. Austin Foundation*), Dr. Thomas Bachmann, John & Caroline Chandler, Stephanie Costelow, Paul Dickson (*Pinola Aviary*), Joan Embery, Jerry Jennings and *The Jennings Foundation*, Kelly Hancock, Ali and Mike Lubbock, Mike McDowell, Fred and Sue Morris, *International Wild Waterfowl Association (IWWA)*, Michael and Judy Steinhardt, John and Judy Todd, and Stephen Wylie. Dr. Pamela Yochem from the *Hubbs-SeaWorld Research Institute* tirelessly kept the financial records organized and the project on budget & schedule.

I am especially indebted to my life-long friends and colleagues Mike & Ali Lubbock (the founders of *Sylvan Heights Bird Park* in Scotland Neck, North Carolina). Virtually the entire staff there aided me greatly over the years, especially Nick Hill, Brent and Katie Gipple Lubbock, Brad and Monica Hazelton, Dustin Foote, John Little, Toad & Hanna Herring, Lissie Glassco, and countless more. Paul Dickson and his curator Jacob Kraemer provided enormous assistance at the *Pinola Conservancy*. In the 1970s and 1980s *SeaWorld of California* (San Diego, California) maintained one of the largest and most comprehensive exotic waterfowl collections in the world. The highly dedicated aviculture staff there, past and present, were always eagerly ready to assist including Scott and Cory Dreischman, Frank Twohy, Stephie Costello, Laurie Burch Conrad, Linda Henry, Paula Hull, Sherry Branch, Laura Whittish, Kim Peterson, and Brad Andrews, among many others. While I was travelling in Europe, Ludger, Eli, Lorenz, and Alexander Bremehr hosted me many times at their most impressive waterfowl farm in Germany, and I also spent much productive waterfowl time with Peter, Coba, and Robert Kooy in Holland. I also thank Gotz Kubler, Chris Marler, and Bill and Fran Makins, among others for their generous hospitality.

I also thank Linda Santos (*Honolulu Zoo*), Glynn Young (*Durrell Wildlife Conservation*, Saint Heller, Jersey) Jeff Sailer and many other private aviculturists who bent over backward to help me, including Arnold and Debbie Schouten,

Maynard Axelson, Mickey and Connie Ollson, Lynn Dye, Paul Dixon, Jacob Kraemer, Ian Gereg, Phil Stanton, Gus and Debbie Ben David, Joe and Sally DeSarro, Nancy Collins, Walt and Gay Sturgeon, Sheila McKay, and Nancy Collins.

An army of waterfowl lovers were literally always ready to lend me a hand, including Susan Adie and Brad Stahl, Jim and Jennine Antrim, Dr. Thomas Bachman, Rangus and Sheila Baird, Lorraine Betts, Pete and Jen Clement, Suzana Machado D'Oliveira, Jack Eitinear, Dick Filby, Bill Frazer and Donna Patterson, David and Christa Kaplan, John Kernan, Denise Landau, Anne Lemenager, Jim and Stephie Mahoney, Bob Meade, Pepe and the Mendez family at *Mundo Marino, Argentina*, Mike McDowell, Buzz and Bev Nason, Ester Pereira and Joe Razim, Betsy Pincherra, Micheline Place, Monica Schillat, Tony Soper, Dr. Brent S. Stewart, Wayne and Sue Trivelpiece, Anna and Alex Vdovenko, Victoria and Charley Wheatley, and Werner and Susan Zehnder.

I also thank Susan Adie, Alicia Berlin, Joan Embrey, Jerry Jennings, Mark Lockwood, Duane Pillsbury, Dr. Brent S. Stewart, Phil Unitt, Kim Urehara, Dr. Pamela K. Yochem and many other fine friends for their constant and enthusiastic support and friendship. Near the end of the project when I was diagnosed with stage 3 melanoma, I fortunately had several first class doctors, and some of the most skilled oncologists in the world, including Dr. Michael Kosty, Dr. Michael Proscino, Dr. Brendan Gaylis, Dr. Ghada Kassab, and Dr. Raymond Press. They kept me going when the picture appeared quite grim.

Though most of the photos used in the book are mine, I was nonetheless helped by a number of especially talented photographers who generously allowed me to use some of theirs, including: Maynard Axelson, James Barnes, Glenn Bartley, Jeff Bernier, Tom Blackman, Mike Coots, Cristiano Crotte, Raul Delgado, Joe and Sally DeSarro, Jack Eitiner, Jonathan Fiely, Ken Fink, Ian Gereg, Richard Grosz, John Haig, Bill Johnson, Bob Kothenbeutel, Gary Kramer, Greg Lavaty, James Lees, Mark Lockwood, Katie Gipple Lubbock, Chris Malachowski, Tony Mercieca, Gerard Monteaux, Michael Morel, Victor Murayama, Judd Patterson, Michael Patrikeev, Gary H. Rosenberg, Debbie Shouton, Dr. Brent S. Stewart, Joshua Stiller, David Stimac, Harold Stiver, Robert Straub, Matthew Studebaker, Walter Sturgeon, Kim Uyehara, and John Webster.

John and Lisa Bass skillfully prepared all of the images and were indispensible in designing the volume, and Suzy and Chris Johnson held down the fort at the home office during my often and long absences in the field.

Black-Bellied Whistling-Duck

WHISTLING-DUCKS

Also known as tree-ducks, whistling-duck is a far more suitable name because several species virtually never perch in trees, and the especially vocal birds are renowned for their loud, repetitious, multi-syllabic whistles that are reminiscent of twittering songbirds. The sexes are similar. Their short, broad, rounded wings have primaries and secondaries of nearly equal length. Flying rather slowly on shallow, leisurely wing-beats, their slightly drooped necks and heads extend well forward. Projecting well beyond the short tail their long legs give flying ducks the impression of having long, pointed tails. Their characteristic flight style and distinctive hunch-back posture is diagnostic. Due to the legs being located closer to the center of the body than most ducks, they have a very erect posture and walk with grace. Primarily vegetarian, they feed mainly on seeds, grain, sedges, grasses, aquatic plants, bulbs, tubers, berries and fruit. Their strong pair-bonds probably extend beyond one season. Whistling-ducks nest in a variety of sites ranging from concealed ground nests to tree cavities or other hollows. Constructed by both pair partners, nests are only scantily lined with down, probably because both sexes incubate, thus eggs are seldom uncovered. Drakes possibly assume the greater role during the 24- to 31-day incubation period. Sharp claws and stiffened tails enable ducklings to scamper out of deep nesting cavities. The striking ducklings are conspicuously marked with contrasting patterns. Both parents are very protective of offspring, even carrying ducklings on their backs while swimming.

BLACK-BELLIED WHISTLING-DUCK
Dendrocygna autumnalis

Length 21" (53 cm); Wingspan 30" (76 cm); 1.8 lbs (0.8 kg)
- Bright waxy-red bill & black belly; *a.k.a.* Red-Billed Whistling-Duck
- Diagnostic white upper-wing patch
- Gray face with white eye-ring
- Pink legs & feet
- More muted juvenile has blue-gray bill, legs & feet
- Population size: 1,000,000 to 2,000,000; increasing

BLACK-BELLIED WHISTLING-DUCK

Juveniles

FULVOUS WHISTLING-DUCK
Dendrocygna bicolor

Length 19" (48.5 cm);
Wingspan 26" (66 cm); 1.5 lbs (0.67 kg)
- Rich tawny head & under-parts, with darker back & wings
- Whitish rump & pale flank stripes conspicuous in flight
- Dark grayish-blue bill, legs & feet
- Juveniles grayer & duller with less distinctive spotting
- Global population size: ca. 1,500,000

Juvenile

10

WEST INDIAN WHISTLING-DUCK
Dendrocygna arborea

Length 18.9-22.8 " (48-58 cm);
Wingspan 35-39" (90-100 cm); 2.5 lbs (1.1 kg)
- Largest, bulkiest Whistling-Duck: *a.k.a.* Cuban or Black-Billed Whistling-Duck
- Mostly dark brown, with mottled back & black-&-white markings on flanks
- Ventral spotting of duller juvenile reduced & more streaked
- Found on most Caribbean islands, but a relatively rare species that is quite vulnerable
- Population size: 10,000 to 20,000; decreasing

Juvenile

Emperor Goose

GEESE

Wild geese on the move are renowned for impressive wedge, or V-formations, and staggered lines that can be surprisingly long. The spectacle of countless skeins of geese streaming across a pale pastel dawn sky, or tens of thousands plunging into roosts as the setting sun casts a magic golden glow over pristine marshes spawn dynamic images forever etched in the memory. Descending geese literally tumble out of the sky from large flocks, when they alternately side-slip from side-to-side downward in steep glides on stiff, set, half-folded wings, skillfully catching themselves with beating wings just prior to touching down. The highly gregarious geese frequently feed and roost in exceedingly large flocks, with some congregating in countless thousands during passage and on staging and wintering grounds. Except for size, the sexes are identical. Well known for strong family ties, goose pairs can remain bonded for life. Millions of geese nest on the tundra and in the vast wetlands of the high Arctic, where an ample food supply, little competition and continuous daylight enables them to feed for up to 20 hours a day. Females alone incubate for 19 to 30 days, while the notoriously aggressive ganders vigorously challenge other geese, and often repel predators as large as foxes. Juveniles are guided to traditional staging and wintering grounds by their parents. Families migrate north together in the spring, but once on the breeding grounds, yearlings separate from the adults.

GREATER WHITE-FRONTED GOOSE
Anser albifrons

Length 28" (71 cm);
Wingspan 4.4-5.1' (1.3-1.6 m); 4.0-7.0 lbs (1.8-3.2 kg)
- White front around bill base extends to dark forehead
- Pink or orange bill with white tip; bright orange legs & feet
- Variable dark blotches on brown belly
- White rump; blackish tail tipped white
- 40-50% have prominent yellow orbital ring
- Juvenile lacks white forehead until well into 1st winter; lacks black belly bars until 1st summer; duller bill & legs

PACIFIC WHITE-FRONTED GOOSE
A. a. sponsa

Western Alaska & winters Pacific coast from Oregon to Mexico
2nd group in the Bristol Bay lowlands & winters in NC Mexico
- Long bill; can show yellow eye-ring
- Overall brownish coloration variable
- Population size: >600,000; increasing

GREATER WHITE-FRONTED GOOSE

Pacific White-Fronted Goose

Immatures

GREATER WHITE-FRONTED GOOSE

TULE GOOSE
A. a. elgasi

SW Alaska; winters Sacramento Valley, Calif.
- 10% larger & 15% longer bill than *A. a frontalis*; ganders to 7.0 lbs (3.3 kg); overall darker brown; long neck
- 40-50% have prominent yellow orbital ring
- Population size: ca 5,000 to 10,000

Juveniles

GREATER WHITE-FRONTED GOOSE

GAMBEL'S WHITE-FRONTED GOOSE
A. a. gambelli
NW Canada & adjacent Alaska (Mackenzie River Basin); winters lower Mississippi Valley to Texas & Mexico
- Large; long pinkish bill; long neck; relatively long legs
- Slightly smaller than similar Tule Goose
- Population size: ca 700,000; stable

GREENLAND WHITE-FRONTED GOOSE
A. a. flavirostris
West Greenland; winters in Ireland & W Scotland
- Darkest race, with variable extensive black belly bars- some geese have nearly completely black under-parts
- Population size: ca. 24,000; fluctuating

Immature

LESSER WHITE-FRONTED GOOSE
Anser erythropus

Length 19-26" (48-66 cm)
Wingspan 4.4' (135 cm); 3.0-5.0 lbs (1.4-2.7 kg)
Eurasia; infrequent North America vagrant
• Small rounded head, with very steep forehead
• Extensive white blaze extends to above eye
• Conspicuous yellow eye-ring; short pink bill
• Global population size: ca. 28,000 to 33,000; declining

Juvenile

LESSER WHITE-FRONTED GOOSE

Juvenile

BEAN GOOSE
Anser fabalis

Length 31" (79 cm);
Wingspan to 5.7' (1.7 m); to 7 lbs (3.2 kg)
Eurasia; vagrant to Alaska & west coast
- Black bill with varying amounts of orange
- Legs & feet bright orange-yellow
- Basically brown overall, with paler grayish-brown breast
- Darker head, neck & wings evident in flight
- In 2007, some authorities split the 5 Bean Geese races into 2 species- Taiga Bean Goose *A. fabalis* & Tundra Bean Goose *A. serrirostris*, each with several races
- Global population size: 100,000 to 120,000; declining

PINK-FOOTED GOOSE
Anser brachyrhychus

Length 22-28" (56-71 cm);
Wingspan 4.6' (140 cm); 4.2-4.9 lbs (1.8-2.4 kg)
- Darker head & neck contrasts with grayish body
- Dark brown neck well furrowed; pink legs & feet
- Black bill with bright pink band behind black nail & variable pink along sides of upper mandible
- Pale bluish-gray forewings very conspicuous in flight
- Population size: >400,000; increasing

Fledgling

GREYLAG GOOSE
Anser anser

Length 33" (85 cm);
Wingspan 5.6' (175 cm); to 8.6 lbs (3.9 kg)
Eurasia; rare vagrant
- Large stout orange-pink bill
- Overall rather uniform gray-brown
- Conspicuous pale-gray forewings
- Legs & feet pink or flesh-pink
- Wild ancestor of most domestic geese
- Global population: >1,000,000

Fledgling

EMPEROR GOOSE
Chen canagica

Length 26" (66 cm); Wingspan 3.9' (1.2 m); 6.1 lbs (2.8 kg)
- Blue-gray plumage appears scaled black & white
- White head & neck often stained rusty by iron deposits in the water
- Small pinkish bill but lower mandible can be black
- Short neck; bright yellow-orange legs & feet, white tail
- Vaguely resembles Blue Goose, but projects cleaner, more delicate appearance
- Juvenile duller, with brown rather than black barring on back; gray mottling on head & fore-neck.
- Global population size: 50,000 to 90,000; fluctuating

EMPEROR GOOSE

SNOW GOOSE
Chen (Anser) caerulescens

Length 28-31" (71-79 cm);
Wingspan 4.5-4.7' (1.3-1.4 m); 5.3 to 10+ lbs (2.4-4.5 kg)
- Overall white with black wing tips
- Deep pink to crimson bill has black 'grinning patch'
- Head & upper neck may be stained orange from iron-rich water
- Legs & feet deep pink
- Juvenile pale gray dorsally, but darker on head & neck; flight feathers dark brown; gray bill, legs & feet
- 2 races & dark color morph (Blue Goose)
- Population size: > 5,000,000; increasing

LESSER SNOW GOOSE
C. c. caerulescens
- Prevails in western part of range

SNOW GOOSE

Juvenile

Immatures

Juvenile

Blue Goose

GREATER SNOW GOOSE
C. c. atlanticus
NE Canada, NW Greenland; winters Atlantic seaboard
- Much larger than nominate race, weighing to 10.5 lbs (4.8 kg), with proportionally heavier, longer head & bill exceeding 2.4" (6 cm)
- Population size: ca 1,000,000; fluctuating

Immature

ROSS'S GOOSE
Chen rossii

Length 23" (58 cm); Wingspan 3.8' (1.2 m); 2.7-4.0 lbs (1.2-1.8 kg)
- Entirely white plumage except for black primaries
- 25-40% smaller than similar Snow Goose, with relatively short neck
- Short reddish bill may have bluish-gray warts at base
- Lacks 'grinning patch' of Snow Goose; feather line at bill base nearly straight, not curving strongly forward
- Juvenile has gray head & neck; stubby gray bill gradually becomes pinkish
- Goslings can be yellow or gray, with a number of varations
- Population size: >1,000,000; increasing

ROSS'S GOOSE

Subadults

BLUE MORPH ROSS'S GOOSE
- Rare blue morph has white face & belly, with dark body & wings; lacks 'grinning patch'; has nearly straight feather line at bill base, & may show bluish-gray warts at bill base
- While Ross's & Snow Geese do hybridize, measurements of some individuals conform exactly to those of *C. rossi*, but most captive birds are probably hybrids

Juveniles

28

CANADA GOOSE
Branta canadensis

Length 25-45" (61-114 cm);
Wingspan 3.6-5.0' (1.1-1.5 m); 6-20 lbs (2.7-9.0 kg)
- Black head & neck; white chinstrap
- Brownish-gray dorsally; white rump & black tail
- Bill, legs & feet black
- Some populations sedentary
- Seven races with considerable overlap

WESTERN CANADA GOOSE
B. c. moffitti
Mainly Great Plains to the southern prairie region of W Canada; winters S & SW U.S.
- Longer, thinner neck, with relatively shorter bill & legs than *B. c. maxima*
- Some may have white collar at base of neck
- Also known Moffitt's or Great Basin Canada Goose
- Population size: >338,000; increasing

CANADA GOOSE

GIANT CANADA GOOSE
B. c. maxima

Mainly eastern Great Plains; essentially resident in southern part of breeding range, but also winters as far south as the Gulf Coast
- Many show conspicuous white spots on forehead that may extend across face
- Global population size: >3,500,000; increasing

Juvenile

ATLANTIC CANADA GOOSE
B. c. canadensis

SE Canada, NE U.S., W Greenland; winters Atlantic coast
- Rarely shows white neck-ring, but creamy-white base of neck, upper breast & extreme upper back forms ill-defined broad collar
- Population size: >80,000-230,000; declining

CANADA GOOSE

DUSKY CANADA GOOSE
B. c. occidentalis
Glacier Bay northward to Cook Inlet; winters BC, Washington & Oregon
- Quite dark overall except for white vent
- Population size: ca. 13,000; fluctuating

Juvenile

VANCOUVER CANADA GOOSE
B. c. fulva
Glacier Bay south to coastal region & islands of BC; southern breeders largely sedentary, but more northerly breeders migrate south to BC, Washington, Oregon & N California
- Very dark chocolate-brown; darkest Canada Goose
- Population size: ca. 220,000

CANADA GOOSE

LESSER CANADA GOOSE
B. c. parvipes
Central Alaska east to NW Hudson Bay & south through the Canadian prairie provinces; winters Oregon & Calif.
- Bill quite short but not stubby
- Ill-defined race linking smaller northern & larger southern forms
- Global population size: ca. 300,000; fluctuating

INTERIOR (HUDSON BAY) CANADA GOOSE
B. c. interior
Manitoba, Ontario & Quebec north to coasts of Hudson Bay; winters mainly in the Mississippi Basin
- Normally smaller & darker overall than nominate race, with somewhat shorter & narrower bill
- Global population size: ca. 1,600,000; stable to increasing

CACKLING GOOSE
Branta hutchinsii

Length 25-27" (64-68 cm);
Wingspan 3.5-3.9' (1.1-1.2 m); 3-5 lbs (1.4-2.5 kg)
- Black head & neck; white chinstrap
- Brownish-gray dorsally; white rump & black tail
- Juvenile duller & less contrasting
- The 4 small tundra-breeding "Canada Goose" forms were split from the larger, more southerly races in 2004

RICHARDSON'S CACKLING GOOSE
B. h. hutchinsii
Arctic Canadian coast & islands east to Baffin Island & W Greenland; winters mainly W. Texas south to north central Mexican highlands; also Lousiana Gulf Coast
- Pale plumage, especially on breast
- Some show a very pale white neck collar
- Short stubby bill
- Population size: ca. 700,000; fluctuating to increasing

TAVERNER'S CACKLING GOOSE
B. h. taverneri
Interior Alaska & NW Canada; winters mainly Washington, Oregon & California
- Highly variable plumage ranges from pale to quite dark
- May have rather broad, narrow or incomplete white neck-ring
- Population size: ca. 40,000; declining

CACKLING GOOSE

CACKLING GOOSE
B. h. minima
W coast Alaska; winters mainly in Washington & Oregon
- Smallest race, size of a large Mallard
- Very dark; round head, short neck; relatively long legs
- Very short stubby bill; may have thin incomplete white neck-ring
- Population size: ca. 300,000; increasing

ALEUTIAN CACKLING GOOSE
B. h. leucopareia
Western Alaska (mainly Aleutian Islands); winters in the Cental Valley Calif.
- Prominent white neck-ring may be quite broad in front; tiny deep-based bill
- Numbered merely 200-300 in 1963; by 2009-79,500
- Population size: ca. 160,000; increasing

HAWAIIAN GOOSE (NENE)
Branta sandvicensis

Length 22-28" (56-71 cm);
Wingspan 4.6' (140 cm); 4.2-4.7 lbs (1.9-2.1 kg)
- Distinct lines of feather grooves on both sides of neck
- Reduced foot webbing facilitates walking on steep lava slopes
- Thin black line at base of neck
- Juvenile has duller body & furrowed neck pattern
- Population size: ca. 2,500

HAWAIIAN GOOSE (NENE)

Juveniles

BARNACLE GOOSE
Branta leucopsis

Length 27" (69 cm);
Wingspan 4.2' (1.3 m); 3.7-5.3 lbs (1.7-2.4 kg)
NE Greenland; vagrant to Atlantic coast
• White face, with black streak extending from bill to dark brown eye
• Bluish-gray, dorsally barred with black; whitish under-parts
• Black neck & breast; bill, legs & feet dark gray
• Population size: 870,000; increasing

Juvenile

BRENT GOOSE
Branta bernicla

Length 25" (63 cm);
Wingspan 3.5' (1.6 m); 3.1 lbs (1.4 kg)
- Black head, neck, back, wings & tail
- White neck-ring & rump; black legs & feet
- Bill quite short but not stubby
- Among most maritime of geese; feeds mainly on eel grass & sea lettuce

BLACK BRANT
(Pacific Brent Goose)
B. b. nigricans
Arctic North America; winters western US & Mexico
- Nearly complete white neck-ring
- Whitish upper flanks contrast with blackish belly
- Population size: ca. 9,000; fluctuating

BRENT GOOSE

Black Brant

ATLANTIC (PALE-BELLIED) BRENT
B. b. hrota
Canada, Greenland; winters east coast of US
- Small neck collar, pale sides & whitish underparts
- Population size: ca. 158,000; increasing

BRENT GOOSE

Atlantic Brent

Juvenile

Subadult

Trumpeter Swan

SWANS

Largest and most majestic of waterfowl, the graceful swans are pure white as adults, whereas juveniles are usually brownish-gray. Male swans are known as cobs, females as pens and downy young as cygnets. While not dimorphic, cobs tend to be larger and more aggressive than their mates. They possess the greatest number of cervical vertebra of any warm-blooded animal. Their 23-25 neck vertebra provide them with incredible neck flexibility, enabling them to feed on the bottom in deep water. Swans spend considerable time in the water, but some also graze ashore. Pairs normally mate for life, so long as both partners are alive, but a new mate will generally ultimately be taken if a partner is lost. Pens are responsible for incubation, but both parents care for broods. Cobs defending nests or cygnets are exceptionally formidable adversaries and fights between rivals can be vicious, sometimes even resulting in the death of the loser. The 4- to 8-ounce cygnets are little more than puffs of down at hatching, and upon advancing through the so-called 'ugly duckling' stage, they metamorphose into the classic form and beauty of adults. The highly migratory Arctic-breeding swans may travel thousands of miles. Normally flying at 35-50 mph, flight speed may exceed 70 mph when assisted by stiff tail winds. Swans fly with their long neck fully extended and legs trailing behind. Juveniles remain with their parents until the onset of the following breeding season, enabling them to learn the well-established migration routes when families wing south in the fall and north in the spring.

TRUMPETER SWAN
Cygnus buccinator

Length 6' (1.8 m);
Wingspan 6.6' (2 m); to 28 lbs (12.7 kg)
- Largest, tallest, most statuesque waterfowl
- Completely white but head & neck often stained from iron deposits in the water
- Gray-brown juvenile has pink bill with black at base & tip
- Population size: >46,000; increasing

TRUMPETER SWAN

- Distinct red stripe ('lipstick line') bisects upper & lower mandible
- Forehead slopes evenly to large, straight, jet-black bill
- Population size: ca. 50,000; only 100 birds existed in the 1930s, reintroductions have increased the population since

Immatures

WHOOPER SWAN
Cygnus cygnus

Length to 6' (1.8 m); Wingspan 6.6' (2 m); to 23.8 lbs (10.7 kg)
Regular Alaska vagrant from NE Siberia & rare breeder on Attu Island
- Entirely white plumage, with black & yellow bill, black legs & feet
- Named for distinctive powerful call
- Grayish juvenile pinkish bill has dark tip
- Global population size: ca 180,000

WHOOPER SWAN

- Bright lemon-yellow bill patch extends from eye to point in front of nostrils; extent & pattern of yellow individually variable

Juveniles

TUNDRA SWAN
Cygnus columbianus

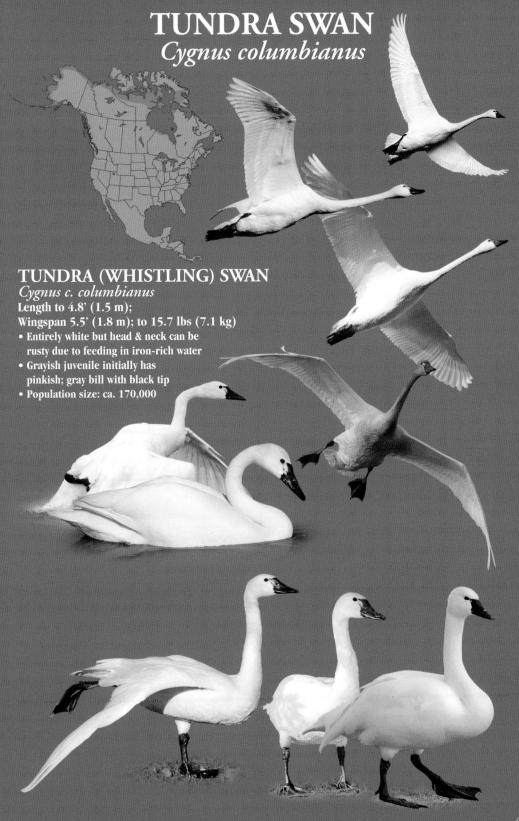

TUNDRA (WHISTLING) SWAN
Cygnus c. columbianus
Length to 4.8' (1.5 m);
Wingspan 5.5' (1.8 m); to 15.7 lbs (7.1 kg)
- Entirely white but head & neck can be rusty due to feeding in iron-rich water
- Grayish juvenile initially has pinkish; gray bill with black tip
- Population size: ca. 170,000

TUNDRA SWAN

- Black bill usually has variable yellow spot in front of eye, but spot can be absent, & some swans can show a slight red 'lipstick line'

Whistling Swan

Juveniles

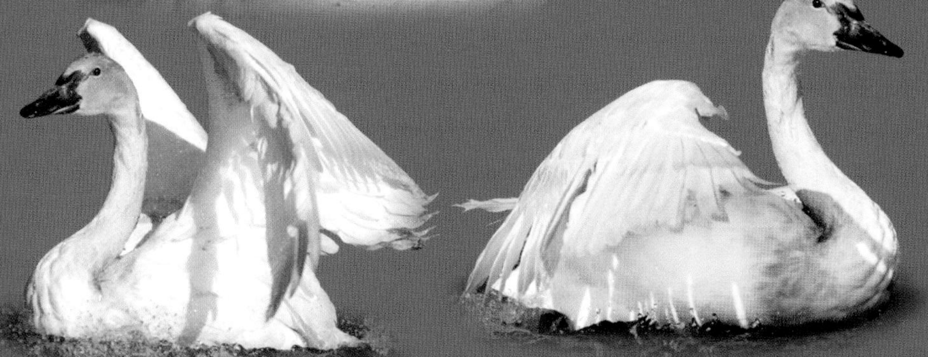

Immatures

47

TUNDRA SWAN

TUNDRA (BEWICK'S) SWAN
Cygnus c. bewickii
Length 4.6' (1.4 m); Wingspan 5.5' (1.8 m); to 15.5 lbs (7 kg)
Eurasia; rare vagrant to Alaska
• Treated as Whistling Swan race but possibly is a distinct species due to the extent of yellow on the bill, smaller size & different range
• Population size: ca 20,000

48

TUNDRA SWAN

- Highly variable bill pattern, with lemon-yellow patch ending bluntly in front of nostrils

Bewick's Swan

Juveniles

Subadult

Immatures

MUTE SWAN
Cygnus olor

Length 5' (1.52 m); Wingspan 6.3' (1.9 m); to >29 lbs (13.5 kg)
- Completely white plumaged; legs & feet dark gray-black; eye pale brown
- Partially leucistic "Polish" Mute Swan has pinkish legs & feet
- Neck often arched rather than held straight; long tail projects beyond the legs in flight
- Large black knob on reddish-orange bill much reduced when not breeding
- Gray-brown juvenile dark gray bill becomes pinkish during 1st winter
- Extremely aggressive during breeding season when cobs can charge with wings held high & arched, a behavior known as *busking*
- Not mute; utters variety of calls, as well as hissing, gurgling & snorting
- Introduced in late 1800s from Europe & considered a pest in some areas; >25,000 from the Great Lakes to Atlantic coast alone, with <700 in the west; control measures have begun

MUTE SWAN

Subadult

Busking ♂s

Polish Mute Swan

Juveniles

Subadults

Wood Duck

PERCHING-DUCKS

Adorned with bright colors and patterns of astonishing complexity and beauty, Wood and Mandarin Duck males are arguably among the most brilliantly plumaged of all birds. The Wood Duck is an American native, whereas the Asian Mandarin is an introduced exotic. Despite their brilliant coloration, due to their broken pattern plumage, drakes can easily melt away into the background in their world of light and shadow. Prime habitat consists of secluded pools, marshes, swamps, lakes and slow-moving or swifter rivers in broad-leafed forests featuring dense tree and shrub cover overhanging the banks. The ducks fly with considerable agility, especially when weaving through thick timber, where their fairly long, broad, square-tipped tails enable them to skillfully maneuver. Wood Ducks easily drop nearly vertically into small woodland ponds. Proportionally larger eyes than other waterfowl facilitate life in the dim light of their shaded haunts, perhaps enabling them to perceive greater detail when adeptly winging though the trees at twilight in near darkness. Most active during the early morning and evening hours, the omnivorous Wood Duck feeds on a wide variety of items, but in the fall and winter, acorns, chestnuts, beechnuts and hickory nuts are particularly favored. Dependent on wooded regions with sufficient water and tree cavities high above the ground for nesting, they can cling to vertical trunks like enormous woodpeckers do, bracing themselves with their long tails. Female Wood Ducks incubate 9-15 eggs for around 40-60 days.

WOOD DUCK
Aix sponsa

Length 18.5" (47 cm); Wingspan 30" (76 cm); 1.3 lbs (0.6 kg)
- Large shaggy mostly green erectile crest; white throat & long tail
- Reddish bill, with black tip; forward half of upper mandible white
- Red eyes with bright orange-red orbital ring; legs & feet orange-yellow
- Brownish-gray female has bushy crest & broad white eye-ring
- Eclipse male retains bright red eye & orbital ring; muted bill colors
- Juvenile male has mostly yellow bill & dull red eyes; lacks red orbital ring
- Population size: 2,800,000; stable or increasing

WOOD DUCK

WOOD DUCK

Subadult ♂s

Eclipse

MANDARIN DUCK
Aix galericulata

SW Russia, NE China & Japan
Length 20" (51 cm); Wingspan 29" (74 cm); 1.5 lbs (0.69 kg)
- Orange-chestnut 'side whiskers'; greatly enlarged inner vane of 12th secondary feather forms spectacular orange-gold 'sail' up the flanks
- Female similar to Wood Duck, but paler, with thin, white eye-ring extending to nape as narrow white line
- Feral population of several hundred breeds in Sonoma County, California
- Global population size: 65,000 and declining

MANDARIN DUCK

MUSCOVY DUCK
Caririna moschata

Length 33" (84 cm); Wingspan 4.9' (150 cm);
Male avg. 6.6 lbs (3 kg); Female avg. 2.8 lbs (1.2 kg)
- Upper-parts strongly glossed with green & purple iridescence; striking white fore-wing & under-wing
- Up to 3 times the size of females, males have a feathered domed crown with erectile hanging crest
- Pale flesh-pink bill, with black nail & black across middle & base of upper mandible
- Blackish, bare swollen skin around eyes joins bill base
- Enlarged blackish to dark red knob extends along top of upper mandible
- Smaller female has no bulge at bill base & may lack bare facial skin
- Domestic drakes can be massive, grotesque multicolored ducks weighing as much as 11 lbs (5 kg), with ugly red facial warts & huge caruncles
- Global population size: ca. 300,000; declining overall though several populations are increasing

MUSCOVY DUCK

♂s

Juvenile

*Ducklings are typically patterened yellow & dark brown, but some are dark variants

Northern Pintail

DABBLING DUCKS

Dabbling or puddle ducks are typically simply known as dabblers due to their distinct "dabbling" feeding style. Rather short legs centrally positioned toward the sides of the body bring about their familiar duck waddle. Swift, agile fliers, a number undertake lengthy migrations. The majority are markedly sociable, gregarious ducks that winter in large, mixed concentrations. They commonly up-end or tip-up to forage on the bottom. In stark contrast to their cryptically patterned mates, most drakes are brilliantly plumaged. However, the colorful plumage is replaced during the post-nuptial summer molt, by an obscure eclipse plumage that resembles their more somber mates. During the summer molt all flight feathers are shed, thus the ducks are flightless for about three weeks. The dull eclipse plumage provides the flightless males with a degree of needed camouflage and is normally worn for two or three months, but longer in some species. In migratory dabbling ducks, monogamous pair bonds are short and seasonal, with new mates courted annually. Displaying can commence as early as late autumn when drakes assume nuptial plumage, but communal courtship is normally initiated in winter flocks. Dabblers typically nest on the ground, usually in thick vegetation, and at times some distance from water. Most species lay relatively large clutches of up to a dozen eggs that are incubated by females for 21-28 days.

GADWALL
Anas strepera

Length 20" (51 cm);
Wingspan 33" (84 cm); 2 lbs (0.9 kg)
- Grayish overall; pale brown head; white belly & black rump
- Both sexes have white wing-patch
- Narrow bill mostly blackish; female bill orange with dusky saddle
- Legs & feet orange-yellow with gray webs
- Global population size: ca. 6,000,000; North American population ca. 3,500,000 and increasing

GADWALL

Juvenile ♂s

Eclipse

Eclipse

AMERICAN WIGEON
Anas americana

Length 20" (51 cm);
Wingspan 32" (81 cm); 1.6 lb (0.7 kg)
- Blackish iridescent green patch through eye; pinkish breast & sides
- Gray head; creamy-white stripe on forehead & crown center; speckled cheeks & throat; black rump & white belly
- Blue-gray bill tipped black; white fore-wing patch; green & black speculum
- Female mainly gray & brown with white belly; head has blackish-brown crown
- Global population size: 2,600,000; stable

AMERICAN WIGEON

EURASIAN WIGEON
Anas penelope

Length 20" (51 cm); Wingspan 32" (81 cm); 1.5 lbs (0.7 kg)
Eurasia; regular NA & Greenland vagrant
- Rufous head, with yellowish-buff crown stripe
- Blue-gray bill with dark tip; white fore-wing patch
- Brownish-rose breast; gray back & sides
- Legs & feet blue-gray; dark brown eye
- Female head, neck & breast variable; cinnamon-buff to rufous-brown, but often much grayer
- Global population size: 3,000,000

EURASIAN WIGEON

Juvenile ♂s

♀

Eclipse

FALCATED DUCK
Anas falcata

Length 18-21" (51 cm); Wingspan 32" (81 cm); 1.3-1.7 lbs (0.7 kg)
Asia; vagrant to Pacific coast, especially Alaska
- Iridescent green & bronzy-purplish head, with long, drooping crest
- Long sickle-shaped wing feathers overhang tail; scaly breast
- Long blackish bill; legs & feet dark gray
- Short, chunky brownish female appears large headed
- Global population size: ca. 80,000 and declining

FALCATED DUCK

♀

Juvenile ♂

Eclipse

Eclipse

68

MALLARD
Anas platyrhynchos

NORTHERN MALLARD
Anas p. platyrhynchos
Length 23" (58 cm);
Wingspan 35" (89 cm); 2.4 lbs (1.1 kg)
- Iridescent green head; white neck-ring; rusty-purple breast
- Drake bill yellow; female bill dusky-brown to yellowish-orange with dark saddle
- Blue speculum with white borders; orange legs & feet
- Curled black central tail feathers
- Legs & feet orange
- Population size: ca. 20,000

Just fledged

GREENLAND MALLARD
Anas p. conboschas
- Paler than nominate race & 10% heavier, with somewhat smaller bill
- Restricted to coastal southern Greenland, where mainly sedentary
- Population size: ca. 15,000 to 30,000 (but recent estimates of ca. 100,000)

MEXICAN DUCK
Anas p. diazi
Extreme SW U.S. & N central Mexico
- Sexes similar; resembles small female Mallard, but darker & richer in color
- Bill olive-yellow; female bill more olive-green, with only traces of orange at base
- Legs & feet deep orange; duller in female
- Greenish-blue speculum less blue than Mallard
- Possibly all north of U.S. border are hybrids (Mexican Duck x Mallard)
- Population size: ca. 55,000 but large fluctuations

MARIANA MALLARD
Anas p. oustaleti
- Formerly regarded as separate species, it probably was an unstable hybrid, with some closely resembling Mallards, whereas others renamed Pacific Black Ducks
- Restricted to the Mariana Islands (Guam, Saipan & Tinian), it became extinct in 1981

AMERICAN BLACK DUCK
Anas rubripes

Length 23" (58 cm);
Wingspan 35" (89 cm); 2.6 lbs (1.2 kg)
- Bill olive-yellow; female bill olive-green with black flecking
- Mainly brownish-black; white under-wing; no white in tail
- Purplish-blue speculum with black borders
- Male legs & feet reddish-orange; female orange-brown
- Population size: ca. 620,000; declining

AMERICAN BLACK DUCK

74

MOTTLED DUCK
Anas fulvigula

Length 22" (56 cm);
Wingspan 30" (76 cm); to 3.1 lbs (1.4 kg)
- Darker, smaller & thinner necked than similar female Mallard
- Black spot at gape of bright yellow bill
- Female bill olive-drab to dull orange, with olive spots concentrated in middle of upper mandible
- Legs & feet orange

GULF MOTTLED DUCK
Anas f. maculosa
- Overall darker & more heavily marked than shorter necked Florida race
- Population size: ca. 600,000

MOTTLED DUCK

FLORIDA MOTTLED DUCK
Anas f. fulvigula
- Slightly longer necked & less heavily marked than the Gulf race, hence appears paler
- Population size: ca. 40,000; stable

76

LAYSAN DUCK
Anas laysanensis

Laysan

Length 16" (40 cm);
Wingspan 30" (72 cm); to 1.5 lbs (0.7 kg)
- Buff to reddish brown with dark brown markings; legs & feet orange
- Extent of white facial patch variable; older ducks may show more white
- Male may have faint green iridescence on head & slightly upturned central tail feathers
- Speculum green in males & dull brown in females & juveniles
- Bill gray-green; female bill brownish-yellow with dull orange on sides
- In 2004 & 2005, 42 were translocated to Midway Atoll NWR, where the ducks are thriving
- Population size: ca. 500 to 700; increasing

LAYSAN DUCK

HAWAIIAN DUCK (KOLOA MAOLI)
Anas wyvilliana

Length 19" (49 cm); Wingspan 31" (79 cm); to 2 lbs (0.9 kg)
Hawaii; Kauai & reintroduced to Oahu & Hawaii
- Variable in overall coloring, with whitish eye-ring
- Male blackish-brown crown; hind-neck has varying amounts of green iridescence
- Male central tail feathers may be slightly up-curved
- Green speculum bordered in front with whitish-buff & behind with black & then white
- Legs & feet orange in both sexes
- Bill olive-gray, with black nail & darker base & center to upper mandible
- Female dusky gray-brown bill, fleshy-yellow behind nail
- Population size: ca. 2,200

HAWAIIAN DUCK (KOLOA MAOLI)

22 Days

10 Days

72 Days

EASTERN SPOT-BILLED DUCK
Anas zonorhyncha

Length 22" (56 cm); Wingspan 30-33" (80 cm); 1.7-2.2 lbs (0.9 kg)
East Russia, Mongolia, N & E China, Korea & Japan; vagrant to Alaska
- Black bill with bright yellow tip
- Buff face with black stripe through eye
- Blue speculum finely bordered with white
- Legs & feet bright orange-red
- Both sexes & juvenile overall buff-brown
- No eclipse plumage
- Population size: 600,000 to 1,000,000; declining

Juvenile

BLUE-WINGED TEAL
Anas discors

Length 15.5" (39 cm);
Wingspan 23" (58 cm); to .95 lbs (0.43 kg)
- Gray-violet head, with bold white facial crescent
- Pale blue forewing patch & metallic-green speculum
- Reddish-buff with dark spotting; white flank patch
- Bill black; legs & feet orange-yellow with dusky webs
- Female brown-gray, with pale area at base of bill
- Population size: 8,000,000 to 9,000,000

BLUE-WINGED TEAL

83

CINNAMON TEAL
Anas cyanoptera

Length 16" (41 cm);
Wingspan 22" (56 cm); 14 oz (397 g)
- Cinnamon overall; reddish-orange eyes
- Both sexes have powder blue shoulder patch & iridescent green & black speculum, but brighter in males
- Female similar to Blue-Winged Teal, but warmer buff with less pronounced pale patch at base of bill, & indistinct eye-line
- Longer, more shoveler-like bill than Blue-Winged Teal
- Juvenile similar to female
- Global population size: 450,000; stable to declining

CINNAMON TEAL

♀

♀ s

Subadult ♂

Juvenile ♂

Eclipse

NORTHERN SHOVELER
Anas clypeata

Length 19" (48 cm);
Wingspan 30" (76 cm); 1.3 lbs (0.6 kg)
- Green head; white chest; rufous flanks; yellow eyes; orange legs & feet
- Black spatulate bill; brown female bill or orange with dark spots
- Powder-blue forewing patch & iridescent green speculum
- White oval patch at sides of rump
- Global population size: 6,500,000 and stable to slightly declining; North American population size ca 4,000,000 to 5,000,000

NORTHERN SHOVELER

♀s

NORTHERN SHOVELER

Juvenile ♂s

Juvenile ♂s

Eclipse

NORTHERN PINTAIL
Anas acuta

Length to 25" (64 cm);
Wingspan 34" (87 cm); 1.8 lbs (0.9 kg)
- Long, slender pointed tail; brown head & hind-neck
- Fine white line extends up sides of long neck; grayish-blue bill
- Speculum varies from brownish or bronze to coppery-green, with pale cinnamon anterior border & white trailing edge
- Plain female pale brown, especially face & neck
- Global population size: ca. 5,400,000; North American population size 3,400,000

NORTHERN PINTAIL

WHITE-CHEEKED PINTAIL
Anas bahamensis

Length 19" (48 cm);
Wingspan 26" (65 cm); 1.4 lbs (0.65 kg)
- Pure white cheeks, throat & upper fore-neck
- Basal half of bluish bill waxy-red; legs & feet dark gray
- Bright iridescent green speculum
- Similar female slightly smaller & duller than male
- Claimed to be widespread and common in the Caribbean and in South America, though no population estimates available

WHITE-CHEEKED PINTAIL

GARGANEY
Anas querquedula

**Length 15.5" (39 cm); Wingspan 24" (61 cm); 13 oz (369 g)
Eurasia; fairly regular North America vagrant**
- Broad white stripe separates dark crown from reddish-brown face
- Silvery upper wing, with dark green, white-bordered speculum
- Gray sides; bill, legs & feet dark gray
- Elongated scapulars blackish, strongly striped white, & drooping over closed wing
- Global population size: 2,000,000; declining

BAIKAL TEAL
Anas formosa

Length to 16.5' (42 cm);
Wingspan 23" (58 cm); to 15.4 oz (436 g)
NE Russia; vagrant to Alaska & western Pacific coast states
- Distinctive intricately patterned head; bill dark gray, legs & feet yellow-gray
- Speculum green & black, bordering in front with chestnut & behind with white
- Elongated scapulars; striped black, white & cinnamon tertials that droop over the closed wing
- Female has well-defined white cheek spot at bill base
- Global population size: 400,000; declining

BAIKAL TEAL

GREEN-WINGED TEAL
Anas carolinesis

Length 14" (36 cm); Wingspan 23" (58 cm); 12 oz (340 g)
- Green speculum bordered by whitish or buffy bar in front
- Conspicuous vertical white stripe in front of flanks
- Unlike European counterpart, upper side of the iridescent green head stripe lacks buff border & has very weak border below
- Smallest American dabbling duck
- Population size: 3,500,000; increasing

GREEN-WINGED TEAL

EURASIAN GREEN-WINGED TEAL
Anas c. crecca

Eurasia, rare Greenland breeder: Vagrant & breeds in the Aleutian Islands*
- Broad iridescent green head stripe bordered with buff, curving from eye to back of shaggy nape
- Long white stripe along the sides
- Green & black speculum bordered fore & aft with white
- Global population size: 3,600,000 to 4,600,000; slightly declining

*Aleutian Green-Winged Teal (*Anas crecca nimia*) recently judged insufficiently different from *A. c. crecca* to be recognized as a valid race

98

Ring-Necked Duck

POCHARDS

Pochards are heavy-bodied diving ducks with relatively large heads, long necks and proportionally large feet. Sexual dimorphism is generally extreme. While shiny metallic speculums are lacking, many species show broad pale or white panels on the upper wings. The legs of most species are located rather far back on the body causing them to be somewhat awkward ashore. All are superior divers that feed extensively underwater. Many are chiefly vegetarian, whereas others are essentially carnivorous, especially when wintering on brackish or salt water. Fairly short wings obligate pochards to patter over the water to achieve flight speed. A number are essentially freshwater ducks, but salt water is not shunned (particularly in winter), and some species are distinctly marine. Many winter on the open ocean in large flocks, when they are very vulnerable to oil spills and other coastal pollution. Monogamous pair bonds of limited seasonal duration are forged. Drakes then desert incubating mates and commonly join other males to molt. Nests are normally constructed over shallow water or concealed in dense cover, usually not far from water. Floating structures in emergent vegetation are more secure from terrestrial predators, and can be further camouflaged by an overhead screening of reeds and grass stems. Females incubate for around 23-29 days. Brood amalgamation occurs in some species.

CANVASBACK
Aythya valisineria

Length 21" (53 cm);
Wingspan 29" (74 cm); 2.7 lbs (1.2 kg)
- Very long, pointed black bill; long, sloping forehead
- Dark rust-red head & neck; black breast, rump & tail
- Whitish back & sides
- Red eyes (female dark brown); legs & feet blue-gray
- Global population size: 780,000; slightly increasing

CANVASBACK

CANVASBACK

REDHEAD
Aythya americana

Length 19" (48cm);
Wingspan 29" (74cm); 2.7 lbs (1.2 kg)
- Reddish head; black lower neck, breast & rump
- Yellow eyes & bluish bill with black tip
- Female brownish-tawny overall; dark brown eyes; indistinct whitish eye-ring & thin stripe behind eye
- Eclipse plumage & 1st winter male overall dull brown, with reddish-brown head
- Population size: 1,200,000; increasing

REDHEAD

COMMON (EURASIAN) POCHARD
Aythya ferina

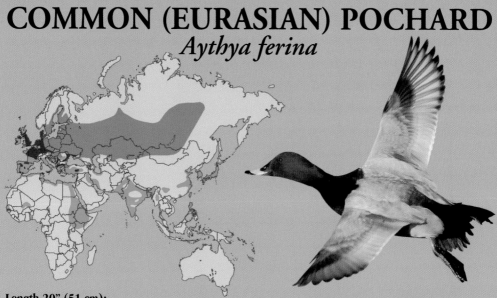

Length 20" (51 cm);
Wingspan 29" (74 cm); 1.8 lbs (0.8 kg)
Eurasia; vagrant, mostly Alaska
- Resembles Redhead but with red eyes
- Glossy chestnut-red head & neck; pale gray body
- Blue-gray bill dark at base & tip; legs & feet gray
- Female eye brown to yellow-brown
- Global population size: 2,200,000; stable or declining

BAER'S POCHARD
Aythya baeri

Length 18" (46 cm);
Wingspan 31" (79 cm); 2.2 lbs (1 kg)
SE Russia; rare vagrant in North America
- Glossy green-black head & upper neck, with small white spot under chin
- Bill bluish-gray; legs & feet gray
- Iris completely white; female eye dark brown
- Global population size: ca. 1,000; declining toward extinction

BAER'S POCHARD

RING-NECKED DUCK
Aythya collaris

Length 17" (43 cm);
Wingspan 25" (64 cm); to 2.0 lbs (0.9 kg)
- Black head & neck has greenish gloss
- Narrow indistinct chestnut neck-ring often barely discernable
- White ring near tip of bluish-gray bill, with thinner white ring at base
- Both sexes have peaked crown; bright yellow male eye contrasts with female brown eye; whitish throat & eye-ring
- Global population size: ca. 1,500,000

RING-NECKED DUCK

RING-NECKED DUCK

TUFTED DUCK
Aythya fuligula

Length 17" (43 cm);
Wingspan 26" (66 cm); 1.6 lbs (0.7 kg)
Eurasia; vagrant to both coasts in North America
- Blackish head & neck glossed purple- may appear greenish
- Long, loose pendent crest, but very short in female
- Dark back & white sides; extensive white upper wing-stripe
- Pale blue-gray bill has black nail; female bill darker
- Both sexes have yellow eyes
- Global population size: 2,600,000; declining

GREATER SCAUP
Aythya marila

Length 18" (46 cm);
Wingspan 32" (81 cm); 2.3 lbs (1.0 kg)
- Black head glossed green; rarely appears purple
- White upper wing-patch extends onto primaries
- Pale blue-gray bill has broad black nail
- Both sexes have yellow eyes; gray legs & feet
- Females show whitish patch at bill base & oval patch by ear
- About 10% larger than similar Lesser Scaup
- Global population size: 1,000,000; declining

GREATER SCAUP

LESSER SCAUP
Aythya affinis

Length 16.5" (42 cm); Wingspan 25" (64 cm); 1.8 lbs (0.8 kg)
- Head glossed purple, but can appear greenish
- Distinct peak at back of crown; black neck & breast
- White upper wing-patch does not extend beyond secondaries, becoming pale brown on primaries
- Both sexes have yellow eyes; gray legs & feet
- Global population size: 4,000,000 to 5,000,000; slightly declining

King Eider

EIDERS

Specialized Arctic-nesting diving ducks, all four Eider species are extremely dimorphic. Nuptial-plumaged drakes show subtle green head pigmentation unique to Eiders. The well-camouflaged female brown coloration is an invaluable asset to ducks requiring concealment on exposed ground nests. Among the most pelagic of waterfowl, non-breeding Eiders are entirely marine, spending virtually all their time at sea. Large numbers winter as far north as possible, wherever open water and ample food is present. The three larger species are ponderous and rise heavily from the water following considerable pattering over the surface. Eiders are quite at ease ashore and walk surprisingly rapidly with an erect, steady rolling gait. Exceedingly proficient divers (few birds dive deeper) with King Eiders reportedly descending a hundred feet or more. Almost exclusively carnivorous, they feed extensively on bivalve mollusks, mussels in particular, that are swallowed whole. Courting drakes utter calls unlike any other waterfowl, and while audible for only a short distance, the unmistakable soft, dove-like cooing of the three larger species assumes a ghostly quality in foggy weather. Pair-bonds are strong initially, but drake interest wanes once egg-laying commences or shortly after, when they depart on the pre-molt migration. Well camouflaged in shallow depressions, nests are heavily lined with copious amounts of the famous Eider down that very effectively insulates the olive-green, bluish or buff eggs. Females rarely depart the nest during the entire 25- 30-day incubation period.

COMMON EIDER
Somateria mollissima

Length 24" (61 cm); Wingspan 38" (97 cm); to 6.2 lbs (2.8 kg)
- Four American races that vary in size, bill shape & overall female color
- Long sloping bill; white head with black cap; green nape
- White back & black flanks, belly & tail; chest pale creamy-pink
- Sides of rump white, extending into large rounded white spot on each side at base of tail; legs & feet grayish
- Females vary from overall gray to rust, with vertically barred flanks; grayer in summer & browner in winter

AMERICAN EIDER *S. m. dresseri*
Breeds from north-east Canada to Massachusetts
- Domed forehead, with large, broadly rounded frontal process
- Bill orangish-yellow to grayish
- Population size: ca. 340,000

COMMON EIDER

American Eider

Juvenile ♂s

Eclipse

COMMON EIDER

NORTHERN EIDER *S. m. borealis*
Breeds Baffin Island & adjacent islands of north-east Canada; Greenland vagrant
- Bill orange-yellow, with variably orange-yellow to olive frontal process
- Population size: 1,400,000; declining

Eclipse

Juvenile ♂

COMMON EIDER

PACIFIC EIDER
S. m. v-nigrum
Breeds Alaskan and Canadian Arctic and Eastern Aleutian Islands
- Especially long sloping head & orange-red bill
- Black V on white throat
- May show white 'sails' on back
- Largest northern duck at 6 lbs
- Population size: 100,000 to 150,000; declining

COMMON EIDER

Juvenile ♂s

Pacific Eider

Eclipse

Eclipse

♀

HUDSON BAY EIDER *S. m. sedentaria*
Resident in Hudson Bay, Canada
- Frontal process less extensive than *S. m. dresseri*
- Female palest of all races
- Population size: 75,000 to 200,000

SPECTACLED EIDER
Somateria fischeri

Length 21" (53 cm); Wingspan 33" (84 cm); 3.4 lbs (1.5 kg)
- Distinctive large silvery-white "goggles" bordered in black; females & ducklings also have prominent spectacles
- Pale green head, with non-erectile shaggy golden-green hood
- Orange bill feathered to nostrils; greenish head & blue eyes
- White upperparts & black breast
- Global population size: 400,000; declining in Alaska

SPECTACLED EIDER

SPECTACLED EIDER

KING EIDER
Somateria spectabilis

Length 22" (56 cm); Wingspan 35" (89 cm); 3.7 lbs (1.7 kg)
- Large, fleshy orange knob atop reddish-pink bill; black V on white throat
- Lime-greenish cheeks; powder-blue crown & nape; white flank spot
- White breast, black back & belly; black wings with white patches
- Both sexes have prominent 'sails' due to scapular feathers protruding from back
- Population size: 800,000 to 900,000

KING EIDER

Displaying ♂

• Female much more reddish than Common Eider, with crescent barring on flanks

♀ s

127

KING EIDER

1st Year ♂s

2nd Year ♂

3rd Year ♂

Eclipse

128

STELLER'S EIDER
Polysticta stelleri

Length 17" (43 cm);
Wingspan 27" (67 cm); 1.9 lbs (0.9 kg)
- White head; greenish nape tuft; green spot in front of eye
- Iridescent bluish or purplish white-bordered speculum
- Black eye-patch, chin & collar; rufous chest with dark spot on side
- Bill, legs & feet blue-gray
- Female dark cinnamon-brown with faint eye-ring; bluish-purple speculum less bright than male
- Superficially more like a dabbling duck than an eider
- Global population size: 110,000 to 125,000; declining

STELLER'S EIDER

Fledgling

STELLER'S EIDER

Juvenile ♂s

Subadult ♂s

Eclipse

Harlequin Ducks

SEA-DUCKS

Most sea-ducks are highly pelagic and well adapted for life in a rugged oceanic environment, even though the majority nest well inland. Sexual dimorphism can be extreme. All are migratory to a greater or lesser degree, but in general they are less apt to undertake long migrations than many waterfowl. While sea-ducks are not especially vociferous, Long-Tailed Duck (Oldsquaw) drakes are renowned for their delightful, far-carrying yodelling calls. Their flight is swift, but most species must laboriously skitter over the water for a considerable distance to gain flight speed. Among the most accomplished of avian divers, Long-Tailed Ducks reportedly descend more than 200 feet, making them the deepest diving waterfowl so far as is known. Primarily diurnal, sea-ducks feed chiefly on animal prey gleaned from the bottom, but the unique mergansers are entirely carnivorous, feeding primarily on fast-swimming fish. Scoters are shellfish specialists that are partial to bivalve molluscs, particularly mussels. Monogamous pair-bonds of seasonal duration are typical, with males forsaking their mates during incubation. Many court communally in winter flocks, often with elaborate and spectacular displays. Clutches of 4-17 buff, olive, green-gray or bluish eggs are incubated by females alone for 24-32 days. Brood amalgamation is not unusual, and ducklings are often essentially independent well before their flight powers are gained.

HARLEQUIN DUCK
Histrionicus histrionicus

Length 16.5" (42 cm);
Wingspan 26" (66 cm); 1.3 lbs (0.6 kg)
- Slate-blue head, breast & back boldly patterned with white
- Large white facial crescent between dark reddish brown eye & bill; deep chestnut flanks
- Short stubby bill, legs & feet blue-gray; long pointed tail
- Dull olive-yellow to blackish-brown female has three white spots on each side of head; brown eyes
- Global population size: 200,000 to 400,000; increasing in North America

HARLEQUIN DUCK

HARLEQUIN DUCK

Diving

135

HARLEQUIN DUCK

Fledglings

Juveniles

Juvenile ♂s

Eclipse

Subadult

LABRADOR DUCK
Camptorhynchus labradorius †

Length 20" (51 cm); Wingspan 30-32" (76-81 cm); 1.1-1.9 lbs (0.5-0.7 kg)
† Extinct 1875; 1st endemic North American bird to disappear
- White head & breast, with black neck ring
- Black cap, upper & underparts; white wings
- Long black bill with swollen orange, yellow or creamy-pink basal half & spoon-like expansion at tip
- Female brownish-gray, with prominent white wing-patch
- Juvenile male apparently like female, but with bit more white on face

♀s

Juvenile ♂

♀

BLACK SCOTER
Melanitta americana

Length 19" (48 cm);
Wingspan 28" (71 cm); 2.1 lbs (1.0 kg)
- Completely black, more glossy on head, neck & upper-parts than under-parts
- Black bill has swollen, orange-yellow cere extending to the nostrils; dark brown eyes
- Upper wings glossy black, but under-sides of flight feathers silvery-gray
- Female dark sooty-brown, with paler cheeks & sides of head; dark olive-brown to blackish bill lacks knob of male, with small amounts of yellow; dark brown eyes
- Juvenile recalls female, but paler on lower head & under-parts Male becomes blacker during 1st winter, but not fully adult until 2nd year
- Male summer plumage duller & browner, with little gloss
- Global population size: 530,000 to 830,000; North America population size 400,000 and slightly increasing

BLACK SCOTER

BLACK SCOTER

1st Year Juvenile ♂s

1st Year Juvenile ♂s

Summer ♂s

COMMON SCOTER

SURF SCOTER
Melanitta perspicillata

Length 20" (51 cm);
Wingspan 30" (76 cm); 2.1 lbs (1.0 kg)
- Black overall, with large white forehead patch & triangular nape patch
- Large heavy colorful bill has swollen base & large black oval on each side
- White eyes; legs & feet orange-red with blackish webs
- Dusky-brown female has dark swollen bill, pale brown eyes; legs & feet dull orange; two whitish patches on side of face; can show white nape patch
- Juvenile male initially like female, but over 1st year becomes mostly black, but highly variable, often with white belly; white forehead patch lacking in 1st year birds
- Population size: 250,000 to 1,300,000; declining

SURF SCOTER

SURF SCOTER

SURF SCOTER

1st Year Juvenile ♂s

1st Year Juvenile ♂s

Full Adult ♂

Summer Plumage ♂

146

WHITE-WINGED SCOTER
Melanitta deglandi

Melanitta deglandi
Length 21" (53 cm);
Wingspan 34" (86 cm); to 4.8 lbs (2.2 kg)
- Overall black, large white wing-patch; juvenile belly sometimes whitish
- Olive-brown flanks can be overlooked; legs & feet red with black webs
- Large orange bill becoming red at tip, with large black knob at base
- White 'comma' below pale blue-gray to white eye
- Sooty-brown female has dark bill, dark eye & two whitish facial spots
- Male summer plumage duller & browner, with little gloss
- Population size: 140,000 to 675,000

Mating Pair

WHITE-WINGED SCOTER

13 Days
26 Days
Juvenile ♂s
Juvenile ♂s

WHITE-WINGED SCOTER

1st Year ♀

1st-2nd Year ♂s

Subadult ♂s

Summer ♂s

VELVET SCOTER

VELVET SCOTER *Melanitta fusca*
Rare vagrant to Greenland
- All black plumage except for conspicuous white wing patch
- Small black basal knob, with sides of upper mandible rich yellow
- Smaller white concentric spot under eye than American race
- Legs & feet orange-red
- Population size: 1,000,000; declining

SIBERIAN SCOTER
Melanitta stejnegeri
Weight to 2.9 lbs (1.3 kg)
Rare Siberian vagrant to Alaska
- Only recently regarded as a full species
- Swollen nasal protuberance extends over nostrils, often forming hooked tip
- Bill has yellow rather than black lamellae
- Mainly black flanks as opposed to brownish
- Global population size: 600,000 to 1,000,000

LONG-TAILED DUCK (OLDSQUAW)
Clangula hyemalis

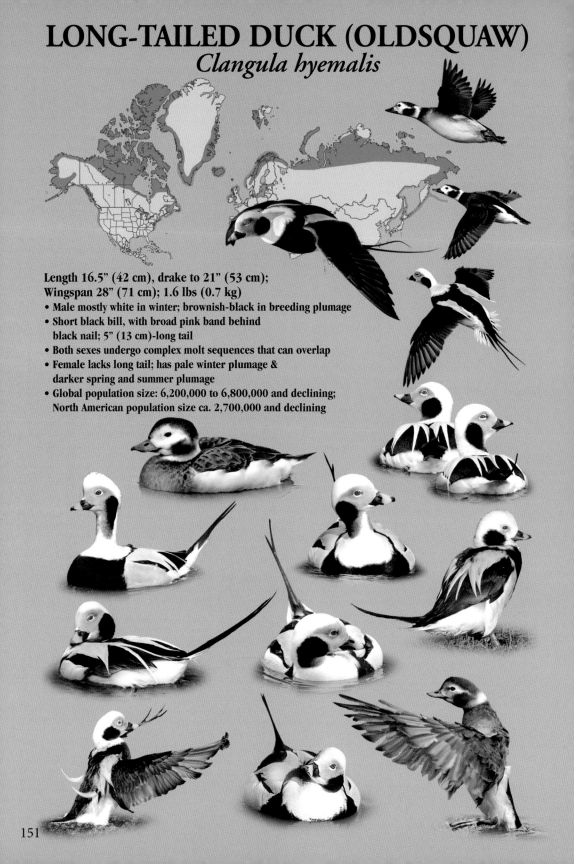

Length 16.5" (42 cm), drake to 21" (53 cm);
Wingspan 28" (71 cm); 1.6 lbs (0.7 kg)
- Male mostly white in winter; brownish-black in breeding plumage
- Short black bill, with broad pink band behind black nail; 5" (13 cm)-long tail
- Both sexes undergo complex molt sequences that can overlap
- Female lacks long tail; has pale winter plumage & darker spring and summer plumage
- Global population size: 6,200,000 to 6,800,000 and declining; North American population size ca. 2,700,000 and declining

LONG-TAILED DUCK (OLDSQUAW)

152

LONG-TAILED DUCK (OLDSQUAW)

BUFFLEHEAD
Bucephala albeola

Length 13.5" (34 cm);
Wingspan 21" (53 cm); 13 oz (369 g)
- High domed head iridescent green, purple & bronze
- Triangular white patch extends from eye to crown
- Black back & under-parts white
- Wing has white speculum, that extends upward to the bend of the wing
- Small blue-gray bill; legs & feet bright pink; dark brown eyes
- Brown-gray female has long narrow small patch below & behind eye; less white on the wing than male
- Population size: 1,000,000; stable or slowing increasing

BUFFLEHEAD

BUFFLEHEAD

Juvenile ♂　　　　　Eclipse

COMMON GOLDENEYE
Bucephala clangula

Length 18.5" (47 cm);
Wingspan 26" (66 cm); 1.9 lbs (0.9 kg)
- Large high-crowned black head glossed green; white round facial spot
- More white on upper wing than larger Barrow's Goldeneye
- Bright yellow eye; female eye paler
- Bill blackish to slate-gray, but female bill can have yellow tip
- Legs & feet yellow-orange
- Global population size: 2,500,000 to 4,000,000; North America population size ca. 1,250,00

COMMON GOLDENEYE

COMMON GOLDENEYE

BARROW'S GOLDENEYE

HOODED MERGANSER
Lophodytes cucullatus

Length 18" (46 cm);
Wingspan 24" (61 cm); to 1.9 lbs (0.9 kg)
- Black head; large white fan crest bordered in black; yellow eyes
- Black back; chestnut sides; white breast has two black bars patch below & behind
- Legs yellow with dark brown areas
- Brownish-gray female has loose, rusty crest & brown eyes
- Population size: 1,100,000; increasing

HOODED MERGANSER

HOODED MERGANSER

SMEW
Mergellus albellus

Length 15" (38 cm);
Wingspan 25" (64 cm); 1.6 lbs (0.7 kg)
Eurasia; regular vagrant to the Aleutian Islands, Alaska, but records from both coasts
- Mostly white; shaggy crest has dark greenish-black stripe
- Elongated erectile feathers on the forehead
- Black patch between eye & relatively short blue-gray bill
- Brown eyes becoming grayish or whitish with age; legs & feet pale pink
- Female grayish overall with head & hindneck dark reddish brown
- Global population size: 130,000; stable or slightly increasing

SMEW

COMMON MERGANSER
Mergus merganser

Length 25" (64 cm);
Wingspan 34" (87 cm); 3.4 lbs (1.5 kg)
- Blackish-green head has slight even crest
- Creamy white breast & flanks; black back
- Both sexes have long, red serrated bill
- Legs & feet deep red; brown eyes
- Reddish-brown female has ragged rufous crest
- Juveniles & eclipse males rather resemble females
- Global population size: 1,700,000 to 2,400,000; increasing

COMMON MERGANSER

COMMON MERGANSER

COMMON MERGANSER

Juvenile ♂s

Juvenile ♂

Juvenile ♂

Eclipse

170

RED-BREASTED MERGANSER
Mergus serrator

Length 23" (58 cm);
Wingspan 30" (76 cm); 2.3 lbs (1.0 kg)
- Black head and ragged crest glossed greenish; red bill & eyes
- Black back; broad white collar; chestnut breast; gray sides
- Drakes remain in full color for very short period of time
- Grayish-brown female has rusty-brown head & shaggy crest
- Global population size: 510,000 to 610,000

RED-BREASTED MERGANSER

RED-BREASTED MERGANSER

173

Ruddy Ducks

STIFF-TAILED DUCKS

Conspicuous rigid tails jauntily cocked up are symbolic of the perky little stiff-tailed ducks. Only two species occur in North America, with the familiar Ruddy Duck the most common, whereas the neotropical Masked Duck appears only periodically, primarily in Florida and Texas. The rather short, broad, basally swollen bill is bright sky-blue in breeding drakes, but fades to brownish-black during the non-breeding season. Among the most aquatic of all wildfowl, stiff-tails are skilled divers with exceptionally large feet. Stiff-tails are among the least vocal of waterfowl but during social displays, drakes are noted for instrumental sounds, such as mechanically generated percussion noises produced by drumming the bill on inflated tracheal air-sacs. Both sexes are quite feisty, threatening with bills agape and readily biting if handled. Courting drakes appear pompous, and their highly visual, stereotyped, complex nuptial antics are most impressive. Rival drakes engage in spirited confrontations, both above and beneath the water, when combatants flail, bite and claw at one another while rising up and leaping around amid much splashing and jockeying for position, but in the end usually only feathers are ruffled. Amorous males can display collectively to seemingly unimpressed females that might even attack courting males. Normal clutches consist of 6-10 eggs, the total weight of which can exceed the weight of laying females. Ducklings require surprisingly little parental care and commonly achieve independence well in advance of fledging when 52-144 days old.

RUDDY DUCK
Oxyura jamaicensis

Length 15" (38 cm);
Wingspan 18.5" (47 cm); 1.2 lbs (0.5 kg)
- Bright chestnut; black crown & nape; white cheek patch (may be spotted)
- Large powder-blue bill; long, stiff tail often elevated
- Can have black spots on white cheeks
- Males rarely become melanistic (black) with age
- Brown female has dark cap & dark line across whitish cheek
- Eclipse male dark gray-brown, with dark cap & dull gray bill
- Global population size: 600,000 and declining; North America population size ca. 485,000 to 500,000

RUDDY DUCK

RUDDY DUCK

MASKED DUCK
Nomonyx dominica

Length 12-14" (30-35 cm);
Wingspan 20-24" (52-62 cm); to 0.84 lbs (0.38 kg)
Neotropical; barely reaches U.S., mostly Florida & Texas Gulf coast, where it has rarely bred
- Head, except nape & lower sides black, forming 'mask'
- Upper-parts, upper breast & flanks rusty-cinnamon
- Under-parts dark brown, mottled with white
- Blackish wings have white patch on outer secondaries
- Bright blue bill, with black tip & nail; legs & feet dark gray
- Female has whitish cheeks & throat, with two dark brown stripes; one through eye & one from base of bill across lower cheeks
- Juvenile similar to female, with green at base of bill
- Global population size: 25,000 to 100,000; fluctuating

MASKED DUCK

MASKED DUCK

MASKED DUCK

Iceland, Reykjavik

URBAN WATERFOWL

One of the most delightful things about urban waterfowl is that they can be virtually fearless. Nearly all cities and a surprising number of small towns maintain city park lakes ranging in size from hundreds of acres to merely a few acres. While countless wild waterfowl take advantage of urban waterways, such aquatic sites are the domain for countless domestic ducks and geese, many of which were released on purpose, including even such large exotics such as Mute and Black Swans. However, most simply appear because town residents commonly release their ducks and geese on a lake to give them a good home. Domestic waterfowl tend to become remarkably tame, with the majority readily feeding from the hand, and in winter this encourages the wild waterfowl to follow suit. For countless urban residents this is really their only exposure to waterfowl and many spend a multitude of pleasant hours every day feeding their ducks. However, for others their initial experiences with city waterfowl becomes a stepping stone to spending quality time with wild birds in the field, where many become naturalists, bird watchers, or avid waterfowl hunters. Thus park domestics are every bit a part of North American waterfowl as the native wild breeders. Most are too heavy to fly but some do, creating the impression that they are wild birds. In addition, escaped or vagrant exotic waterfowl often gravitate to these urban waterways, thus one never knows what might appear in a city park lake on any given day.

BLACK SWAN
Cygnus atratus

Length 4.5' (55");
Wingspan to 6.6' (200 cm); to 13.8 lbs (6.3 kg)
Australia; common in city parks
- Sooty-black, with white primaries & secondaries; black legs & feet
- Bill orange-red to bright waxy-red with white sub-terminal bar & nail
- Bare red skin extends back to reddish or whitish eye
- Peculiar broadened, crinkle-edged greater wing coverts give a ruffled appearance to the top of closed wing
- Neck longer in proportion to body than any other swan
- Juvenile grey-brown, with pale or whitish feather tips producing mottled appearance; gray bill gradually becomes pink & then red
- Global population size: >500,000

SWAN GOOSE
Anser cygnoid

Length 1.0 m (94 cm); Wingspan 6' (185 cm); 7.7 lbs (3.4 kg)
Asia
- Long necked; very long black bill; distinct sloping forehead
- White line around base of bill variable in thickness
- Legs & feet reddish-orange
- Domestic varieties common in city parks
- Global population size: 100,000 to 120,000; declining

BAR-HEADED GOOSE
Anser indicus

Length 30" (76 cm);
Wingspan 5' (1.6m); to 7 lbs (3.1 kg)
Central Asia
- Overall gray; bill yellow-orange
- Broad black stripe over top of head, running eye to eye, & 2nd narrower black stripe parallel & to rear of 1st stripe
- Legs & feet yellow-orange
- Global population size: ca. 63,000; stable or increasing

RED-BREASTED GOOSE
Branta ruficollis

Length to 23" (58 cm); Wingspan to 4.4' (1.3 m); to 3 lbs (1.4 kg)
Siberia; all North America records probably represent escapes
- Smallest, most colorful goose; boldly patterned black, white & chestnut-red; very small stubby black bill; legs & feet black
- Appears dark in flight when bright head & breast often not discernable
- Difficult to census due to shifting wintering locations, hence counts of 88,425 in 2000 vs. 56,860 in 2010 do not necessarily represent a population decrease

RED-BREASTED GOOSE

EGYPTIAN GOOSE
Alopochen aegyptiacus

Length 29" (73 cm);
Wingspan 5' (154 cm); to 5.4 lbs (2.4 kg)
Africa; feral in many North America sites; breeds in southeast Florida & possibly California
- Conspicuous dark brown eye-patch
- Rather long pink bill with black nail, nostrils & mandible edges
- Pink legs & feet may be tinged purple
- Primaries strongly glossed with green; large white upper wing patch
- Sexes similar but female smaller
- Global population size: 200,000 to 500,000

COMMON SHELDUCK
Tadorna tadorna

Length 26" (67 cm);
Wingspan 4.3' (133 cm); 2.6 lbs (1.2 kg)
Eurasia; most North America sightings are escapes, but several Newfoundland records could represent wild birds
- Breeding male has bright carmine-red bill, with prominent fleshy knob
- Black head & upper neck glossed green; bronze-green speculum
- Legs & feet dark flesh-pink
- Smaller, duller female lacks fleshy bill knob & may show variable whitish area around bill base
- Global population size: 580,000 to 710,000

RUDDY SHELDUCK
Tadorna ferruginea

Length 26" (67 cm);
Wingspan 4.7' (145 cm); 3 lbs (1.4 kg)
Eurasia; most records are escapes,
but the species is regarded
as hypothetical; has been
recorded in Greenland
- Overall rusty-orange, with paler buff head
- Rump, tail & upper tail coverts black, lightly glossed green
- Bill, legs & feet blackish
- Narrow black collar around base of neck lacking in non-breeders & females
- Global population size: 100,000 to 200,000

FERRUGINOUS DUCK
Aythya nyroca

Length 16" (42 cm);
Wingspan 23" to 28" (60 to 70 cm);
1.1 to 1.3 lbs (520 to 580 gms)
Central Europe through Eurasia to western Mongolia and NW and central China; scattered North Africa, Turkey, Iran, Pakistan, Aghanistan, Kasmir; escapees occasional in North America
• Short beak, head appears triangular
• Males deep russet brown, white triangular patch under tail, white belly
• Broad white wing bar in all birds only visible in flight
• Females and juveniles similar to male but dull brown and no reddish color
• White iris in adult male and brown in adult female
• Global population size: 160,000 to 260,000; fluctuating

RED-CRESTED POCHARD
Netta rufina

Length 21-22" (53-57 cm);
Wingspan 33-35" (84-88 cm); 2.6 lbs (1.2 kg)
Eurasia
- Rich chestnut head with busy crest
- Bill bright vermillion has pink nail
- Red eyes; legs & feet orange-red
- Primaries & secondaries white with brown
- Brown female has very pale cheeks, throat & upper neck

Appendix

TAXA	BODY MASS	EGGS	INCUBATION	FLEDGING
Whistling Ducks				
Black-Bellied Whistling-Duck *Dendrocygna autumnalis*	Both sexes; 1.6-2.2 lbs (avg 1.8 lbs)	12-16 buffy-white eggs	26-31 days (avg. 27)	53-63 days
Fulvous Whistling-Duck *Dendrocygna bicolor*	Both sexes; avg. 1.5 lbs	9-15 cream to buff-white eggs	24-28 days	55-63 days
West Indian Whistling-Duck *Dendrocygna arborea*	Both sexes; avg. 2.5 lbs	8-12 (avg. 9) milky white eggs	30 days	50 – 60 days
Geese				
Greater White-Fronted Goose *Anser albifrons*	M avg. 5.0 lbs; F avg. 4.5 lbs	4-7 light-buff, pale pinkish-white or cream colored eggs	24-28 days	40-65 days
Pacific White-Fronted Goose *Anser a. sponsa*	M avg. 5.3 lbs; F avg. 4.9 lbs	4-7 light-buff, pale pinkish-white or cream colored eggs	24-28 days	40-65 days
Tule Goose *Anser a. elgasi*	M avg. 6.7 lbs; F avg. 6.3 lbs	4-7 light-buff, pale pinkish-white or cream colored eggs	24-28 days	40-65 days
Gambel's White-Fronted Goose *Anser a. gambelli*	M avg. 7.0 lbs; F avg. 6.7 lbs	4-7 light-buff, pale pinkish-white or cream colored eggs	24-28 days	40-65 days
Greenland White-Fronted Goose *Anser a. flavirostris*	M avg. 5.6 lbs; F avg. 5.7 lbs	4-7 light-buff, pale pinkish-white or cream colored eggs	24-28 days	40-65 days
Lesser White-Fronted Goose *Anser erythropus*	M 4.3-5.0 lbs; F 3.0-4.7lbs	4-6 light yellowish-white eggs	25-28 days	35-40 days
Bean Goose *Anser fabalis*	M 5.9-7.0 lbs; F 5.2-6.3 lbs	3-7 pale straw-colored eggs Tundra breeders avg. 3-5 eggs Taiga breeders avg. 4-6 eggs	25-30 days. Avg. 25 days in tundra breeders and 28-29 days in taiga breeders	45-50 days
Pink-Footed Goose *Anser brachyrhychus*	M avg. 5.8 lbs; F avg. 5.2 lbs	4-7 white or pale straw-colored eggs	25-28 days	45-50 days
Greylag Goose *Anser anser*	M 7.6-7.9 lbs avg.; F 6.5-6.9 lbs avg.	4-8 creamy-white eggs	28-30 days	55-60 days
Emperor Goose *Chen canagica*	M avg. 6.2 lbs; F avg. 6.1 lbs	4-6 creamy-white eggs	24-27 days	48-55 days
Snow Goose *Chen (Anser) caerulescens*	M 6.0-7.9 lbs; F 5.5-6.7 lbs	4-5 creamy-white eggs	22-25 days	42-50 days
Ross's Goose *Chen rossii*	M avg. 2.9 lbs; F avg. 2.7 lbs	4-5 white (pinkish when just laid) eggs	19-25 days (avg. 22)	40-45 days

Species	Weight	Eggs	Incubation	Fledging
Western Canada Goose *Branta c. moffitti*	M avg. 9.5 lbs; F avg. 8.0 lbs	2-12 dull white or creamy-white eggs. Clutches normally smaller at higher latitudes (avg. 3-5 eggs) than in southern races (avg. 3-9 eggs)	24-28 days. High latitude races incubate for shortest period.	40-60 days
Giant Canada Goose *Branta c. maxima*	M avg. 14.4 lbs; F avg. 12.1 lbs	2-12 dull white or creamy-white eggs. Clutches normally smaller at higher latitudes (avg. 3-5 eggs) than in southern races (avg. 3-9 eggs)	24-28 days. High latitude races incubate for shortest period.	40-60 days
Atlantic Canada Goose *Branta c. canadensis*	M avg. 8.4 lbs; F avg. 7.3 lbs	2-12 dull white or creamy-white eggs. Clutches normally smaller at higher latitudes (avg. 3-5 eggs) than in southern races (avg. 3-9 eggs)	24-28 days. High latitude races incubate for shortest period.	40-60 days
Dusky Canada Goose *Branta c. occidentalis*	M avg. 8.3 lbs; F avg. 6.9 lbs	2-12 dull white or creamy-white eggs. Clutches normally smaller at higher latitudes (avg. 3-5 eggs) than in southern races (avg. 3-9 eggs)	24-28 days. High latitude races incubate for shortest period.	40-60 days
Vancouver Canada Goose *Branta c. fulva*	M avg. 10.2 lbs; F avg. 7.7 lbs	2-12 dull white or creamy-white eggs. Clutches normally smaller at higher latitudes (avg. 3-5 eggs) than in southern races (avg. 3-9 eggs)	24-28 days. High latitude races incubate for shortest period.	40-60 days
Lesser Canada Goose *Branta c. parvipes*	M avg. 6.0 lbs; F avg. 5.4 lbs	2-12 dull white or creamy-white eggs. Clutches normally smaller at higher latitudes (avg. 3-5 eggs) than in southern races (avg. 3-9 eggs)	24-28 days. High latitude races incubate for shortest period.	40-60 days
Interior (Hudson Bay) Canada Goose *Branta c. interior*	M avg. 9.2 lbs; F avg. 8.5 lbs	2-12 dull white or creamy-white eggs. Clutches normally smaller at higher latitudes (avg. 3-5 eggs) than in southern races (avg. 3-9 eggs)	24-28 days. High latitude races incubate for shortest period.	40-60 days
Richardson's Cackling Goose *Branta h. hutchinsii*	M avg. 4.4 lbs; F avg. 4.1 lbs	2-12 dull white or creamy-white eggs. Clutches normally smaller at higher latitudes (avg. 3-5 eggs) than in southern races (avg. 3-9 eggs)	24-28 days. High latitude races incubate for shortest period.	40-60 days
Taverner's Cackling Goose *Branta h. taverneri*	M avg. 4.9 lbs; F avg. 4.5 lbs	2-12 dull white or creamy-white eggs. Clutches normally smaller at higher latitudes (avg. 3-5 eggs) than in southern races (avg. 3-9 eggs)	24-28 days. High latitude races incubate for shortest period.	40-60 days
Cackling Goose *Branta h. minima*	M avg. 4.2 lbs; F avg. 3.0 lbs	2-12 dull white or creamy-white eggs. Clutches normally smaller at higher latitudes (avg. 3-5 eggs) than in southern races (avg. 3-9 eggs)	24-28 days. High latitude races incubate for shortest period.	40-60 days
Aleutian Cackling Goose *Branta h. leucopareia*	M avg. 4.3 lbs; F avg. 4.2 lbs	2-12 dull white or creamy-white eggs. Clutches normally smaller at higher latitudes (avg. 3-5 eggs) than in southern races (avg. 3-9 eggs)	24-28 days. High latitude races incubate for shortest period.	40-60 days
Hawaiian Goose (Nene) *Branta sandvicensis*	M avg. 4.7 lbs; F avg. 4.2 lbs	3-6 white eggs	29-30 days	70-84 days
Barnacle Goose *Branta leucopsis*	M avg. 4.0 lbs; F avg. 3.5 lbs	3-5 grayish-white eggs	24-26 days	40-45 days
Brent Goose *Branta bernicla*	Both sexes avg. 3.1 lbs	3-5 dull creamy white, yellow-white, green-white, or pale-olive eggs	22-28 days (avg. 24)	40 days
Black Brant (Pacific Brent Goose) *Branta b. nigricans*	M. avg. 3.3 lbs; F avg. 3.0 lbs	3-5 dull creamy white, yellow-white, green-white, or pale-olive eggs	22-28 days (avg. 24)	40 days

Atlantic (Pale-Bellied) Brent Goose *Branta b. hrota*	M avg. 3.4 lbs; F avg. 2.8 lbs	3-5 dull creamy white, yellow-white, green-white, or pale-olive eggs	22-28 days (avg. 24)	40 days

Swans

Trumpeter Swan *Cygnus buccinator*	M avg. 26.2 lbs; F avg 20.7 lbs	4-9 dull creamy-white eggs	32-40 days	84-120 days
Whooper Swan *Cygnus Cygnus*	M avg. 23.8 lbs; F avg. 17.9 lbs	4-9 creamy-white or yellow-tinged eggs	Avg. 35 days	70-90 days (avg. 87)
Tundra (Whistling) Swan *Cygnus c. columbianus*	M avg. 15.7 lbs; F avg. 13.7 lbs	2-7 (avg. 4-5) creamy or dull white eggs	30-40 days (avg. 32)	60-70 days
Tundra (Bewick's) Swan *Cygnus c. bewickii*	M: avg. 14.1 lbs; F avg. 12.5 lbs	4-6 creamy-white or slightly yellowish eggs	29-30 days	50-70 days, but perhaps as few as 40-45 days
Mute Swan *Cygnus olor*	M avg. 26.9 lbs; F avg. 19.6 lbs	1-11 (avg. 4-6) dull gray, greenish-white or pale blue-green eggs	35-38 days	120-150 days

Perching-Ducks

Wood Duck *Aix sponsa*	M avg. 1.5 lbs; F avg. 1.2 lbs	9-15 dull white to brownish-white eggs	25-31 days	60 days
Mandarin Duck *Aix galericulata*	M avg. 1.4 lbs; F avg. 1.1 lbs	9-15 pure white oval eggs	28-30 days	40-45 days
Muscovy Duck *Caririna moschata*	M avg. 6.6 lbs; F avg. 2.8 lbs	8-15 creamy-white eggs	35 days	Perhaps as long as 120 days

Dabbling-Ducks

Gadwall *Anas strepera*	M avg. 2.2 lbs; F avg. 1.9 lbs	5-15 (avg. 8-10) dull cream-white to pale pinkish white blunt, oval eggs	24-28 days	45-63 days
American Wigeon *Anas americana*	M avg. 1.7 lbs; F avg. 1.5 lbs	6-12 (avg. 7-9) creamy-white eggs	Avg. 25 days	37-63 days (avg. 40-50)
Eurasian Wigeon *Anas penelope*	M avg. 1.6 lbs; F avg. 1.4 lbs	6-16 (avg. 8-9) cream-white or pale-buff oval eggs	24-25 days	40-50 days
Falcated Duck *Anas falcata*	M avg. 1.7 lbs; F avg. 1.3 lbs	6-9 cream-white or yellowish eggs	24-26 days	45-60 days
Mallard *Anas platyrhynchos*	M avg. 2.1-2.6 lbs; F avg. 1.8-2.3 lbs	8-13 light gray-green, buff, with or bluish blunt, oval eggs	26-30 days	49-60 days
Mexican Duck *Anas p. diazi*	Averages 10% larger than nominate race	8-13 light gray-green, buff, with or bluish blunt, oval eggs	26-30 days	49-60 days
Mariana Mallard *Anas p. oustaleti*	Slightly smaller than nominate race	8-13 light gray-green, buff, with or bluish blunt, oval eggs	26-30 days	49-60 days

American Black Duck *Anas rubripes*	M avg. 2.7 lbs; F avg. 2.5 lbs	7-12 cream-white to green-buff eggs	26-29 days	49-56 days
Gulf Mottled Duck *Anas fluvigula maculosa*	M avg. 2.7 lbs; F avg. 2.6 lbs	7-12 cream-white to green-buff eggs	26-29 days	49-56 days
Florida Mottled Duck *Anas f. fluvigula*	M avg. 2.7 lbs; F avg. 2.6 lbs	7-12 cream-white to green-buff eggs	26-29 days	49-56 days
Laysan Duck *Anas laysanensis*	M avg. 15.7 ounces; F avg. 15.9 ounces	4-8 greenish-white eggs	26 days	50-60 days
Hawaiian Duck (Koloa Maoli) *Anas wyvilliana*	M avg. 1.5 lbs; F avg. 1.3 lbs	6-13 (avg. 8-10) greenish-white eggs	26-28 days	50-60 days
Eastern Spot-Billed Duck *Anas zonorhyncha*	M 2.2 – 3.3 lbs; F avg. 1.7 – 3.0 lbs	6-14 white grayish or greenish eggs	ca. 28 days	49-56 days
Blue-Winged Teal *Anas discors*	M avg. 14.1 ounces; F avg. 13.1 ounces	8-13 olive-white to creamy-white eggs	21-27 days (23-24 avg.)	35-44 days
Cinnamon Teal *Anas cyanoptera*	M avg. 14.4 ounces; F avg. 1.7 ounces	9-12 buff-pinkish or pure white eggs	21-25 days	35-49 days
Northern Shoveler *Anas clypeata*	M avg. 1.4 lbs; F avg. 1.3 lbs	9-11 pale olive-buff to green-gray oval eggs	22-28 days (avg. 25)	40 to 60 days
Northern Pintail *Anas acuta*	M avg. 1.8 lbs; F avg. 1.7 lbs	6-10 greenish oval eggs	21-26 days	38-52 days
White-Cheeked Pintail *Anas bahamensis*	M 1.0-1.2 lbs; F 1.1-1.4 lbs	5-12 buff eggs	25 days	44-60 days
Garganey *Anas querquedula*	M avg. 13.9 ounces; F avg. 13.1 ounces	8-11 light-brown or pale straw-colored oval eggs	21-25 days	35-40 days
Baikal Teal *Anas formosa*	M avg. 15.4 ounces; F avg. 15.2 ounces	6-9 pale-greenish eggs	24-25 days	45-55 days
Green-Winged Teal *Anas carolinensis*	M 11.0-14.0 ounces	8-11 dull yellowish-white, or pale olive-buff, blunt oval eggs	21-23 days	25-44 days (avg. 40)
Eurasian Green-Winged Teal *Anas c. crecca*	M avg. 15.1 ounces; F avg. 13.9 ounces	8-11 dull yellowish-white, or pale olive-buff, blunt oval eggs	21-23 days	25-44 days (avg. 40)

Pochards

Canvasback *Aythya valisineria*	M avg. 2.8 lbs; F avg. 2.5 lbs	8-10 bright gray-olive eggs	23-29 days	60-77 days
Redhead *Aythya americana*	M avg. 2.4 lbs; F avg. 2.4 lbs	7-16 pale-buff to greenish eggs	23-29 days	55-75 days
Common (Eurasian) Pochard *Aythya ferina*	M avg. 2.1 lbs; F avg. 1.7 lbs	6-10 green-gray broad, oval eggs	24-28 days	50-55 days

Species	Weight	Eggs	Incubation	Fledging
Baer's Pochard *Aythya baeri*	M avg. 1.9 lbs; F avg. 1.5 lbs	6-10 yellowish-gray, cream or pale-brown eggs	ca. 27 days	50-60 days
Ring-Necked Duck *Aythya collaris*	M avg. 1.7 lbs; F avg. 1.5 lbs	5-14 eggs (avg. 8-10) olive-gray, buff-green, olive-brown, pale-cream or buffy-brown eggs	25-29 days	49-56 days
Tufted Duck *Aythya fuligula*	M avg. 1.7 lbs; F avg. 1.6 lbs	6-14 (avg. 10) yellow-brown to olive-gray oval eggs	23-28 days	45-50 days
Greater Scaup *Aythya marila*	M avg. 2.2-2.3 lbs; F avg. 2.0-2.3 lbs	8-11 large olive-gray blunt, oval eggs	24-28 days	35-45 days
Lesser Scaup *Aythya affinis*	M avg. 1.9 lbs; avg. 1.7 lbs	8-10 dark olive-buff eggs	24-28 days	45-50 days

Eiders

Species	Weight	Eggs	Incubation	Fledging
Common Eider *Somateria mollissima*	M avg. 5.0 lbs; F avg. 4.5 lbs	3-6 lusterless pale brown, olive-green, greenish-gray or rarely bluish blunt, oval eggs	25-30 days	65-75 days
American Eider *Somateria m. dresseri*	M avg. 4.4 lbs; F avg. 3.3 lbs	3-6 lusterless pale brown, olive-green, greenish-gray or rarely bluish blunt, oval eggs	25-30 days	65-75 days
Northern Eider *Somateria m. borealis*	M avg. 4.4 lbs; F avg. 4.0 lbs	3-6 lusterless pale brown, olive-green, greenish-gray or rarely bluish blunt, oval eggs	25-30 days	65-75 days
Pacific Eider *Somateria m. v-nigrum*	M avg. 5.7 lbs; F avg. 5.5 lbs	3-6 lusterless pale brown, olive-green, greenish-gray or rarely bluish blunt, oval eggs	25-30 days	65-75 days
Hudson Bay Eider *Somateria m. sedentaria*	M avg. 5.5 lbs; F avg. 3.7-4.0 lbs	3-6 lusterless pale brown, olive-green, greenish-gray or rarely bluish blunt, oval eggs	25-30 days	65-75 days
Spectacled Eider *Somateria fischeri*	M avg. 3.6 lbs; F avg. 3.2 lbs	3-6 lusterless pale brown, olive-green, greenish-gray or rarely bluish blunt, oval eggs	25-30 days	65-75 days
King Eider *Somateria spectabilis*	M avg. 3.8 lbs; F avg. 3.6 lbs	3-6 lusterless pale brown, olive-green, greenish-gray or rarely bluish blunt, oval eggs	25-30 days	65-75 days
Steller's Eider *Polysticta stelleri*	Both sexes; 1.8-1.9 lbs	7-9 olive-buff or olive-brown eggs	ca. 26 days	ca. 45-55 days

Sea-Ducks

Species	Weight	Eggs	Incubation	Fledging
Harlequin Duck *Histrionicus histrionicus*	M avg. 1.5 lbs; F avg. 1.2 lbs	4-8 creamy-yellow or light-buff blunt, oval eggs	27-30 days	56-70 days
Labrador Duck *Camptorhynchus labradorius*	M avg. 1.9 lbs; F avg. 1.1 lbs	Undescribed	Undetermined	Undetermined
Black Scoter *Melanitta americana*	M avg. 2.4 lbs; F avg. 1.8 lbs	6-11 (avg. 6-8) cream, light-buff or pinkish-buff oval eggs	25-31 days	45-50 days
Common Scoter *Melanitta nigra*	M avg. 2.3 lbs; F avg. 2.1 lbs	6-11 (avg. 6-8) cream, light-buff or pinkish-buff oval eggs	25-31 days	45-50 days

Species	Weight	Eggs	Incubation	Fledging
Surf Scoter *Melanitta perspicillata*	M avg. 2.2 lbs; F avg. 2.0 lbs	5-9 non-glossy, buffy-white or pinkish or pale-buffy oval eggs	25-31 days	50-67 days
White-Winged Scoter *Melanitta deglandi*	M avg. 3.4 lbs; F avg. 2.7 lbs	5-17 (avg. 7-9) creamy, pinkish or pale-buffy oval eggs	25-31 days	50-67 days
Velvet Scoter *Melanitta fusca*	M avg. 3.8 lbs; F avg. 3.6 lbs	5-17 (avg. 7-9) creamy, pinkish or pale-buffy oval eggs	25-31 days	50-67 days
Siberian Scoter *Mellanitta stejnegeri*	M avg. 3.4 lbs; F avg. 2.7 lbs	5-17 (avg. 7-9) creamy, pinkish or pale-buffy oval eggs	25-31 days	50-67 days
Long-Tailed Duck (Oldsquaw) *Clangula hyemalis*	M avg. 1.8 lbs; F avg. 1.5 lbs	6-9 cream, yellow-buff or deep olive-buff oval eggs	24-29 days	35-40 days
Bufflehead *Bucephala albeola*	M avg. 15.8 ounces; F avg. 11.4 ounces	6-11 ivory-yellow to pale olive-buff eggs	29-31 days	50-55 days
Common Goldeneye *Bucephala clangula*	M avg. 2.2 lbs; F avg. 1.7 lbs	8-11 clear, pale green,, bluish-green or grayish olivaceous-green blunt, oval eggs	27-32 days	55-68 days
Barrow's Goldeneye *Bucephala islandica*	M avg. 2.4 lbs; F avg. 1.8 lbs	8-11 bluish-green oval eggs	28-32 days	56 days
Hooded Merganser *Lophodytes cucullatus*	M avg. 1.5 lbs; F avg. 1.2 lbs	8-15 pure white (sometimes tinged black) eggs	28-30 days	71 days
Smew *Mergellus albellus*	M avg. 1.6 lbs; F avg. 1.2 lbs	6-10 cream to pale-bluff oval eggs	26-28 days	ca. 70 days
Common Merganser *Mergus merganser*	M avg. 3.7 lbs; F avg. 3.1 lbs	8-15 very pale-buff, ivory-white or yellow broad, oval eggs	30-35 days	60-70 days
Red-Breasted Merganser *Mergus serrator*	M avg. 2.6 lbs; F avg. 2.1 lbs	4-14 (avg. 8-10) deep-buff or green-buff blunt oval eggs	29-35 days	60-70 days

Stiff-tailed Ducks

Species	Weight	Eggs	Incubation	Fledging
Ruddy Duck *Oxyura jamaicensis*	M avg. 1.3 lbs; F avg. 1.1 lbs	6-10 dull creamy-white chalky eggs	20-26 (avg. 23-24) days	52-66 days
Masked Duck *Nomonyx dominica*	M avg. 5.3 lbs; F avg. 3.5 lbs	1-3 pale greenish-white or light buff eggs	24-26 days	90-144 days

Index

Aix galericulata 56, 57
Aix sponsa 53
Aleutian Cackling Goose 34
Alopochen aegyptiacus 188
American Black Duck 73, 74
American Eider 118, 119
American Wigeon 63, 64
Anas acuta 89, 90
Anas americana 63, 64
Anas bahamensis 91, 92
Anas carolinensis 96
Anas clypeata 86
Anas crecca 96
Anas crecca crecca 98, 99
Anas cyanoptera 84, 85
Anas discors 82
Anas falcata 67, 68
Anas formosa 94, 95
Anas fulvigula 75, 76
Anas fulvigula fulviglua 76
Anas fluvigula maculosa 75
Anas laysanensis 77, 78
Anas penelope 65, 66
Anas platyrhynchos 69, 70
Anas platyrhynchos conboschas 71
Anas platyrhynchos diazi 72
Anas platyrhynchos oustaleeti 72
Anas platyrhynchos platyrhynchos 69, 70, 71
Anas querquedula 93
Anas rubripes 73
Anas strepera 61, 62
Anas wyvilliana 79
Anas zonorhyncha 81
Anser albifrons 13
Anser albifrons elgasi 15
Anser albifrons flavirostris 16
Anser albifrons gambelli 15
Anser albifrons sponsa 13
Anser anser 21
Anser brachyrhychus 20
Anser cygnoid 184
Anser erythropus 17
Anser fabalis 19
Anser indicus 185
Anser serrirostris 19
Asian Mandarin Duck 52
Atlantic Pale-Bellied Brent Goose 39, 40
Atlantic Canada Goose 30
Aythya affinis 115, 116

Aythya americana 103, 104
Aythya baeri 106, 107
Aythya collaris 108, 109, 110
Aythya ferina 105
Aythya fuligula 111, 112
Aythya marila 113, 114
Aythya nyroca 191
Aythya valisineria 100, 101, 102
Baikal Teal 94, 95
Bar-Headed Goose 185
Barnacle Goose 37
Barrow's Goldeneye 160, 161
Bean Goose 19
Baer's Pochard 106, 107
Bewick's Swan 48, 49
Black-Bellied Whistling-Duck 7, 8, 9
Black Brant (Pacific Brent Goose) 38
Black Scoter 138, 139, 140
Black Swan 183
Blue Goose 25, 26
Blue-Winged Teal 82, 83
Branta bernicla 38
Branta bernicla brota 39, 40
Branta bernicla nigricans 38
Branta canadensis 28, 29
Branta canadensis canadensis 30
Branta canadensis fulva 31
Branta canadensis interior 32
Branta canadensis maxima 30
Branta canadensis moffitti 29
Branta canadensis occidentalis 31
Branta canadensis parvipes 32
Branta hutchinsii 33
Branta hutchinsii hutchinsii 33
Branta hutchinsii leucopareia 34
Branta hutchinsii minima 34
Branta hutchinsii taverneri 33
Branta ruficollis 186, 187
Branta sandvicensis 35, 36
Branta leucopsis 37
Brent Goose 38, 39, 40
Bucephala albeola 154, 155, 156
Bucephala clangula 157, 158, 159
Bucephala islandica 160, 161
Bufflehead 154, 155, 156
Cackling Goose 33, 34
Camptorhynchus labradorius 137
Canada Goose 29, 30, 31, 32
Canvasback 100, 101, 102

Caririna moschata 58
Chen caerulescens 24, 25, 26
Chen canagica 22, 23
Chen chen atlanticus 26
Chen chen caerulescens 24, 25, 26
Chen rossii 27, 28
Cinnamon Teal 84, 85
Clangula hyemalis 151, 152, 153
Common Eider 118, 119, 120, 121, 122
Common Goldeneye 157, 158, 159
Common Merganser 167, 168, 169, 170
Common (Eurasian) Pochard 105
Common Scoter 141, 142
Common Shelduck 189
Cygnus atratus 183
Cygnus buccinators 42, 43
Cygnus columbianus 46, 47
Cygnus cygnus 44, 45
Cygnus cygnus bewickii 48, 49
Cygnus olor 50
Dabbling-Ducks 60
Dendrocygna arborea 11
Dendrocygna autumnalis 8, 9
Dendrocygna bicolor 10
Dusky Canada Goose 31
Eastern Spot-Billed Duck 81
Egyptian Goose 188
Emperor Goose 12, 22, 23
Eurasian Green-Winged Teal 98
Eurasian Wigeon 65, 66
Falcated Duck 67, 68
Ferruginous Duck 191
Florida Mottled Duck 76
Fulvous Whistling-Duck 10
Gadwall 61, 62
Gambel's White-Fronted Goose 16
Garganey 93
Giant Canada Goose 30
Greater Scaup 113, 114
Greater Snow Goose 26
Greater White-Fronted Goose 13, 14, 15, 16
Greenland Mallard 71
Greenland White-Fronted Goose 16
Green-Winged Teal 96, 97
Greylag Goose 21
Gulf Mottled Duck 75
Harlequin Duck 132, 133, 134, 135, 136
Hawaiian Duck (Koloa Maoli) 79, 80
Hawaiian Goose (Nene) 35, 36
Histrionicus histrionicus 133, 134, 135
Hooded Merganser 162, 163, 164
Hudson Bay Eider 122
Interior (Hudson Bay) Canada Goose 32

King Eider 117, 126, 127, 128
Labrador Duck 137
Laysan Duck 77, 78
Lesser Canada Goose 32
Lesser Scaup 115, 116
Lesser Snow Goose 24, 25
Lesser White-Fronted Goose 17, 18
Long-Tailed Duck (Oldsquaw) 151, 152, 153
Lophodytes cucullatus 162
Mallard 69, 70, 71, 72
Mandarin Duck 56
Mariana Mallard 72
Masked Duck 178, 179, 180, 181
Melanitta americana 138, 138, 140
Melanitta deglandi 147, 148, 149
Melanitta fusca 150
Melanitta nigra 141, 142
Melanitta perspicillata 143, 144, 145, 146
Mellanitta stejnegeri 150
Mergellus albellus 165, 166
Mergus merganser 167, 168, 169, 170
Mergus serrator 171, 172, 173
Mexican Duck 72
Mottled Duck 75, 76
Muscovy Duck 57, 58
Mute Swan 50, 51
Nomonyx dominica 178, 179, 180, 181
Northern Eider 120
Northern Mallard 68, 69, 70, 71
Northern Pintail 60, 89, 90
Northern Shoveler 86, 87, 88
Oldsquaw 131, 150, 151, 152, 153
Oxyura jamaicensis 175
Pacific Eider 121, 122
Pacific Brent Goose 38
Pacific White-Fronted Goose 13, 14. 15
Perching-Ducks 52
Pink-Footed Goose 20
Pochards 99
Polish Mute Swan 51
Polysticta stelleri 129, 130, 131
Puddle Ducks 60
Red-Breasted Goose 186, 187
Red-Breasted Merganser 171, 172, 173
Red-Crested Pochard 192
Redhead 103, 104
Richardson's Cackling Goose 33
Ring-Necked Duck 99, 108, 109, 110
Ross's Goose 27
Ruddy Duck 174, 175, 176, 177
Ruddy Shelduck 190
Sea-Ducks 132
Siberian Scoter 150

Smew 165, 166
Snow Goose 24, 25, 26
Somateria fischeri 123
Somateria mollissima 118
Somateria mollissima borealis 120
Somateria mollissima dresseri 118
Somateria mollissima sedentaria 122
Somateria mollissima v-nigrum 121
Somateria spectabilis 126
Spectacled Eider 123, 124, 125
Steller's Eider 129, 130, 131
Stiff-Tailed Ducks 174
Surf Scoter 143, 144, 145, 146
Swan Goose 184
Swans 41
Tadorna tadorna 189
Tadorna ferruginea 190

Taiga Bean Goose 19
Taverner's Cackling Goose 33
Trumpeter Swan 41, 42, 43
Tufted Duck 111, 112
Tule Goose 15
Tundra Swan 46, 47, 48, 49
Urban Waterfowl 182
Vancouver Canada Goose 31
Velvet Scoter 150
Western Canada Goose 29
West Indian Whistling-Duck 11
Whistling Duck 7
Whistling Swan 46, 47
White-Cheeked Pintail 91, 92
White-Winged Scoter 147, 148
Whooper Swan 44, 45
Wood Duck 52, 53, 54, 55